Diss. ETH No. 16623

Tropospheric transport of water vapour: Lagrangian and Eulerian perspectives

A dissertation submitted to the
SWISS FEDERAL INSTITUTE OF TECHNOLOGY
ZURICH

for the degree of
Doctor of Sciences

presented by
HARALD SODEMANN
Dipl. Geoökologe (Univ. Bayreuth)
B. Sc. (Hons) (Univ. Cape Town)
born 16 December 1975, Esslingen am Neckar
citizen of Germany

accepted on the recommendation of
Prof. Dr. H. C. Davies, examiner
Prof. Dr. H. Wernli, co-examiner
Dr. C. B. Schwierz, co-examiner
Dr. V. Masson-Delmotte, co-examiner

2006

Bibliografische Information der Deutschen Nationalbibliothek

Die Deutsche Nationalbibliothek verzeichnet diese Publikation in der Deutschen Nationalbibliografie; detaillierte bibliografische Daten sind im Internet über http://dnb.d-nb.de abrufbar.

ISBN 3-8325-1384-1

Logos Verlag Berlin
Comeniushof, Gubener Str. 47,
10243 Berlin
Tel.: +49 030 42 85 10 90
Fax: +49 030 42 85 10 92
INTERNET: http://www.logos-verlag.de

Für meinen Vater
und meine Mutter

Contents

Abstract VII

Zusammenfassung IX

1 Introduction 1

1.1 Eulerian tagging approaches . . . 3

1.2 Lagrangian moisture diagnostics . . . 4

1.3 Aims and objectives of this study . . . 4

2 Stable Isotopes of Water 7

2.1 Background and definitions . . . 7

2.1.1 Equilibrium fractionation . . . 8

2.1.2 Rayleigh distillation . . . 10

2.1.3 Non-equilibrium (kinetic) fractionation . . . 12

2.2 Isotope fractionation processes . . . 13

2.3 Isotopic processes in weather systems . . . 16

2.3.1 Mid-latitude cyclones . . . 16

2.3.2 Convective systems . . . 18

2.3.3 Further remarks . . . 19

2.4 Isotopic processes on the seasonal to annual scale . . . 20

2.4.1 Annual mean isotopic signature . . . 20

2.4.2 Stable isotope effects . . . 22

2.5 Isotopic processes on the inter-annual to climate scale . . . 25

2.6 Modelling of stable water isotopes . . . 29

I Lagrangian approach 35

3 A Lagrangian Moisture Source Diagnostic 37
3.1 Identification of moisture uptake 38
3.2 Moisture source attribution 40
3.3 Setup of the calculations 43
3.4 Method validation 45
3.5 Moisture source regions 50
3.5.1 Source regions for Greenland 50
3.5.2 Influence of NAO variability 54
3.5.3 Transport and uptake processes 56
3.5.4 Pre-uptake locations 59
3.6 Further discussion and conclusions 60
3.6.1 Method discussion 60
3.6.2 Moisture transport analyis 63

4 Lagrangian Isotope Modelling 65
4.1 Isotopic fractionation parameters 65
4.1.1 Isotopic fractionation from diagnosed parameters 67
4.1.2 Initialisation with GCM isotope data 68
4.2 Moisture uptake conditions 69
4.2.1 Diagnosed evaporation conditions 69
4.2.2 Diagnosed isotopic composition 71
4.3 Condensation onset conditions 72
4.4 Arrival conditions 75
4.4.1 Wet and dry condition temperatures 77
4.5 Isotopic fractionation modelling 77
4.5.1 Initialisation with GCM output 77
4.5.2 Initialisation with surface parameters 80
4.5.3 Comparison with observational data 80
4.5.4 Assessment of the MCIM isotope model 82
4.6 Relative importance of fractionation temperatures 83
4.7 Further issues 85

4.7.1 The isotope-temperature relationship 85
4.7.2 Transport influences on kinetic fractionation 87
4.8 Concluding remarks 87

II Eulerian approach **91**

5 A Water Vapour Tagging (WVT) Methodology **93**
5.1 Tagging numerics 94
5.1.1 Basic numerics 94
5.1.2 Conservation of mass 96
5.1.3 Positive definite non-oscillatory schemes 97
5.1.4 Explicit numerical diffusion 98
5.1.5 Advection tests of the numerical schemes 99
5.1.6 Tagging algorithms 102
5.2 Idealised 1D advection experiments 105
5.2.1 Model environment and setup 105
5.2.2 Evaluation criteria 107
5.3 Tracer advection experiments 108
5.3.1 1D-Tagging experiments using a uniform wind field 108
5.3.2 1D-Tagging experiments using an oscillating wind field 115
5.4 Concluding assessment of the tagging algorithms 118

6 Implementation of WVT in the CHRM Model **121**
6.1 The CHRM model 122
6.1.1 Code organisation and changes 123
6.1.2 Tracer initialisations 123
6.1.3 WVT output 129
6.1.4 GRIB output 129
6.2 Large-scale dynamics 130
6.2.1 Horizontal and vertical advection 131
6.2.2 Horizontal diffusion 134
6.2.3 Vertical diffusion 136
6.2.4 Lateral boundary relaxation 137

6.3 Phase change parametrisations 138
6.3.1 Condensation and evaporation 139
6.3.2 Grid-scale precipitation 141
6.3.3 Convective precipitation 143
6.3.4 Soil model 146

7 Tagging Study of the August 2002 Flood in Central Europe 147
7.1 CHRM simulation of the event 148
7.2 WVT experiment setup 154
7.3 Tagging simulations 155
7.3.1 Tagged moisture transport 155
7.3.2 Tagged precipitation 161
7.3.3 Consistency of the method 162
7.4 Discussion and further remarks 165
7.4.1 The CHRM tagging implementation 165
7.4.2 The case study of the Elbe flood 166

8 Final Remarks 169

A The transport history of two Saharan dust events archived in an Alpine ice core 177
A.1 Introduction 178
A.2 Data 180
A.2.1 Ice core site and analysis 180
A.2.2 Precipitation data 181
A.2.3 Satellite imagery 182
A.2.4 Meteorological data 182
A.3 Ice core chemistry of dust events 183
A.4 Identification of the dust transport history 185
A.4.1 Back-trajectory calculations 185
A.4.2 Dust deposition 186
A.4.3 Dust mobilisation 187
A.5 Meteorological development during dust transport events 188
A.5.1 Meteorological development during the October dust event . . . 189

A.5.2 Meteorological development during the March dust event 192
A.5.3 Large-scale flow imprints on dust transport 196
A.6 Contributions to the chemical signal in the ice core 200
A.6.1 Dust source region signal 200
A.6.2 Uptake of chemical species during transport 201
A.7 Discussion of the applied methodology 205
A.8 Conclusions 207

B Lagrangian moisture source diagnostic details 209
B.1 Trajectory calculation details 209
B.2 Diagnosed evaporation areas 209
B.3 Uptakes above the boundary layer 209
B.4 Regionalised moisture source regions 212

C MCIM model setup and adjustments 215
C.1 The Mixed-Cloud Isotope Model MCIM 215
C.2 Adjusted lapse-rate calculation 216

Acknowledgments 231

Curriculum vitae 233

Abstract

The hydrological cycle is a key element of the global climate system. Although the atmosphere contains only a small amount of the global water, it takes a crucial role in coupling the major reservoirs, namely inland and sea ice, the oceans, lakes, soils, and rivers via moisture transport and precipitation. Human societies depend on reliable water resources, and are adapted to the present-day hydrological cycle, in particular to the precipitation regime. Understanding the processes that govern moisture transport in the troposphere is therefore of fundamental importance in a changing climate. Identifying the evaporative source regions and transport paths of moisture for particular precipitation events holds the promise of inferring governing mechanisms.

The focus of this work is on enhancing the understanding of the atmospheric branch of the hydrological cycle on a global and regional scale. Adopting a Lagrangian and an Eulerian perspective, two new methods are developed for determining the source regions of precipitation. By means of the two methods, precipitation origin and transport are studied for two target areas at different temporal and spatial scales. The utilised data sources comprise ECMWF's analysis and ERA40 reanalysis data.

The Lagrangian method diagnoses the sources and transport paths of water vapour from 3-dimensional kinematic back-trajectories. The method considers the full transport history of an air parcel. By taking precipitation and subsequent uptakes *en route* into account, each sources' contribution to a diagnosed precipitation at the arrival location can be determined. A Lagrangian analysis of the large-scale moisture transport to Greenland, and its variability with the North Atlantic Oscillation (NAO) during selected winter months reveals the North Atlantic as Greenland's sole moisture source. North Atlantic source regions of moisture are found to vary strongly with the NAO. Moisture sources shift from the Denmark Strait and Norwegian Sea during NAO positive months to the south-eastern and western North Atlantic for NAO negative months. This finding is a new aspect of the influence that large-scale climate modes can impose on the hydrological regime in the target area.

Stable water isotopes in precipitation are employed for a tentative validation of the Lagrangian method. For this purpose, data from the ECHAM4 isotope general circulation model (GCM) and seasonal stable isotope data from a central Greenland ice core site are used. A Rayleigh-type stable isotope fractionation model (MCIM) is applied to predict the isotopic composition of precipitation corresponding to the diagnosed moisture transport conditions, which allows one to compare the model results to observations. Modelled stable isotope ratios show a significant lack of depletion compared to ice core data. This is probably due to the tuning of the MCIM model to different moisture transport conditions. The variability of modelled stable isotope ratios with NAO is however similar to observations from ice-core data for three winter seasons. The variability with the NAO can be qualitatively explained by combined source and transport influences, which indicates the need for a multi-causal interpretation of stable isotope signals in ice cores. The high spatial resolution of the results may be helpful for determining ice-core sites on the Greenland plateau with pronounced NAO variability.

The Eulerian method consists of a water vapour tagging (WVT) implementation into CHRM, a well-verified regional climate model (RCM). Numerical consistency, initialisation and spin-up, and boundary treatment proved to be critical points of the WVT implementation and application. The method is used for determining the sources of precipitation during the Elbe flood (10–13 August 2002), the most severe flood in central Europe in recent decades. A 72 h simulation of the flood period highlighted the importance of the concurrent upper-level circulation for producing extreme precipitation. Identification of the role of the most influential evaporative moisture sources was limited by spin-up effects. However, different evaporative sources did contribute to the extreme precipitation in the most affected area; notably at distinct, subsequent periods of time.

A comparison of the Lagrangian and the Eulerian methods suggests that each approach has different scales and problem settings to which it should preferentially be applied. When problems on the regional scale are considered, or parametrised processes, in particular convection, are important, an Eulerian approach should give the better results. When large-scale transport and processes clearly dominate over parametrised processes, the Lagrangian diagnostic can reveal moisture origins without being limited by a specific RCM domain and spin-up. In this respect, the two methods could be applied jointly, and thereby provide a complementary Lagrangian and Eulerian picture.

With the Lagrangian and Eulerian methods developed in this work, new insight into the characteristics of and the processes related to the atmospheric branch of the hydrological cycle is now attainable.

Zusammenfassung

Der Wasserkreislauf ist ein Schlüsselelement des globalen Klimasystems. Obwohl die Atmosphäre nur einen kleinen Anteil des verfügbaren Wassers enthält, ist sie, mittels Feuchtetransport und Niederschlag, von zentraler Bedeutung für den Austausch zwischen den grossen Wasserspeichern Ozean, Inland- und Meereseis, Seen, Böden und Flüssen. Die Menschheit ist abhängig von verlässlichen Wasserressourcen, und hat sich an den gegenwärtigen Wasserkreislauf und insbesondere die Niederschlagsverteilung angepasst. Angesichts eines sich ändernden Klimas ist daher ein Verständnis der regulierenden Prozesse des atmosphärischen Feuchtetransports unerlässlich. Eine Identifizierung der Verdunstungsquellregionen und Feuchtetransportwege für bestimmte Niederschlagsereignisse verspricht, die zugrundeliegenden Mechanismen zu erschliessen.

Diese Arbeit zielt auf ein erweitertes Verständnis des atmosphärischen Wasserkreislaufs auf globalem und regionalem Massstab ab. Dazu werden zwei Methoden zur Bestimmung der Quellregionen von Niederschlag entwickelt, eine aus Lagrange'scher, die andere aus Eulerscher Sichtweise. Niederschlagsherkunft und -transport werden mit Hilfe der beiden Methoden auf unterschiedlichen räumlichen und zeitlichen Skalen untersucht. Als Datenquellen werden dabei die Analysen und ERA40 Reanalysen des EZMW herangezogen.

Die Lagrange'sche Methode diagnostiziert die Quellen und Transportwege von Wasserdampf aus 3-dimensionalen kinematischen Rückwärtstrajektorien, und berücksichtigt dabei die gesamte Transportgeschichte eines Luftpakets, um die Beiträge einzelner Quellregionen zum Niederschlag am Ankunftspunkt zu bestimmen. Bei der Untersuchung des grossräumigen Feuchtetransports nach Grönland und dessen Schwankungen in Abhängigkeit von der Nordatlantischen Oszillation (NAO) während ausgewählter Wintermonate stellt sich der Nordatlantik als für Grönland alleinig bedeutsame Feuchtequelle heraus. Die nordatlantischen Quellregionen variieren stark mit der NAO; mit dem Wechsel von der positiven zur negativen Phase der NAO verschieben sie sich von der Dänemarkstrasse und dem Norwegischen Meer in den südöstlichen und den westlichen Nordatlantik. Dieses Ergebnis stellt einen neuen Aspekt des Einflusses grossräumiger Klimamodi auf das hydrologische Regime im Untersuchungsgebiet dar.

Stabile Isotope des Wassers im Niederschlag werden für eine erste Validierung der Lagrange'schen Methode eingesetzt. Hierzu werden Simulationsdaten des ECHAM4 Isotopen-Klimamodells und saisonale Isotopendaten aus Eisbohrkernen in Zentral-Grönland verwendet. Die modellierten stabilen Isotopenverhältnisse des Wassers, berechnet mit dem Rayleigh-Fraktionierungsmodell MCIM entsprechend der diagnostizierten Transportbedingungen, ermöglicht einen Vergleich der Ergebnisse der Feuchtetransportdiagnostik mit Messdaten. Im Vergleich mit Eisbohrkerndaten von Zentral-Grönland für drei Winter zeigt sich, dass die Absolutwerte der modellierten stabilen Isotopenverhltnisse deutlich zu wenig abgereichert sind. Dies liegt wahrscheinlich zum Teil im internen Tuning des MCIM-Modells begründet. Die interanuelle Variabilität mit der NAO ist allerdings im Vergleich von Modell und Beobachtungen sehr ähnlich. Die hohe räumliche Auflösung des Grönland-Plateaus in den vorliegenden Ergebnissen kann für die Auswahl von Eisbohrkernen mit ausgeprägten NAO-Signaturen von Nutzen sein. Weiterhin kann die NAO-Variabilität der stabilen Isotopenverhältnisse qualitativ als ein kombinierter Einfluss der Quell- und Transportbedingungen erklärt werden. Dies zeigt die Notwendigkeit auf, diese Einflüsse bei der Interpretation von stabilen Isotopen aus Eisborkernen zu berücksichtigen.

Die Eulersche Methode besteht aus einer Implementierung von markiertem Wasserdampf (MWD) in das vielfach validierte regionale Klimamodell (RKM) CHRM. Bei der Implementierung von MWD erwiesen sich numerische Konsistenz, Initialisierung und Vorlaufzeit, sowie die Behandlung der Modellränder als kritische Bereiche. Die Methode wird angewandt, um die Niederschlagsquellregionen der Elbeflut (10.–13. August 2002), einer der schwersten europäischen Flutkatastrophen der letzten Jahrzehnte, zu bestimmen. Eine 72 h-Simulation der Flutperiode mit MWD stellt die Bedeutung der Strömung in der oberen Troposphäre zu dieser Zeit für das Auftreten der Extremniederschläge heraus. Die Identifizierung der einflussreichsten Verdunstungsregionen ist durch Vorlaufzeiteffekte eingeschränkt. Dennoch können die Beiträge unterschiedlicher Quellregionen zu dem Extremniederschlagsereignis während aufeinanderfolgender Zeitperioden nachgewiesen werden.

Ein Vergleich des Lagrange'schen und des Eulerschen Ansatzes lässt folgern, dass beide Methoden bevorzugt auf bestimmten Skalen und für bestimmte Fragestellungen eingesetzt werden sollten. Werden regionale Massstäbe betrachtet oder sind parametrisierte Prozesse, insbesondere Konvektion, von Bedeutung, so wird die Eulersche MWD Methode die besseren Ergebnisse liefern. Dominieren aber grossräumige Prozesse und Ferntransport, so sollte die Lagrange'sche Diagnostik in der Lage sein, Feuchteherkunft ohne Vorlaufeffekte oder Begrenzung auf ein Modellgebiet zu bestimmen. Bei gemeinsamer Anwendung der beiden Methoden könnte sich so ein Gesamtbild aus Lagrange'scher und Eulerscher Perspektive ergeben.

Mit den beiden in dieser Arbeit entwickelten Methoden sind nun neue Einsichten in die Eigenschaften und Prozesse des atmosphärischen Wasserkreislaufs in greifbare Nähe gerückt.

Chapter 1

Introduction

The hydrological cycle is a key element of the global climate system. Although the atmosphere contains only a small amount of the global water, it takes a crucial role in coupling the major reservoirs, namely the oceans, inland and sea ice, lakes, soils, and rivers, via moisture transport and precipitation (Houghton et al. 2001). Human societies depend on reliable water resources, and are adapted to the present day hydrological cycle, in particular the precipitation regime. Extremes in precipitation, such as drought and flood events, threaten with the loss of life and property. It is therefore vitally important to understand the processes that govern moisture transport in the troposphere, which is even more the case in a changing climate (Christensen and Christensen 2003; Schär et al. 2004).

While it is well known that for example precipitation extremes such as droughts and floods can both be related to remote sea surface temperature (SST) anomalies (Hoerling and Kumar 2003; Rouault et al. 2003), the particular mechanisms and feedbacks remain largely hypothetical. Identifying the evaporative source regions and transport paths for particular precipitation events holds the promise of inferring the underlying mechanisms. However, many moisture sources can potentially contribute to one specific precipitation event, and mixing during transport makes it even more difficult to establish, for example, source-receptor relationships for atmospheric moisture transport (James et al. 2004).

Large-scale climate modes, such as the North Atlantic Oscillation (NAO, Walker and Bliss 1928) can also profoundly alter atmospheric water transport on inter-annual to multi-decadal time scales. The regime changes of such climate modes are incorporated into a multitude of data archives, such as ice cores from Greenland (e.g. Appenzeller et al. 1998b; Vinther et al. 2003), and hence can serve as proxy records of their past behaviour (Hurrell et al. 2003). The variability of the general circulation associated with such climate modes in turn offers the opportunity, for example by means of atmospheric reanalysis data, to study key influences on the present-day atmospheric water cycle.

Diagnoses of the present-day atmospheric water budget in a defined region were first conducted by bulk methods, based on wind and station data (Budyko 1974; Trenberth 1997). Robasky and Bromwich (1994) in addition used radiosonde data to derive a precipitation estimate for Greenland from a budget approach. Moisture transport has often been inferred from the examination of circulation patterns and instantaneous moisture fluxes at different vertical levels (e.g. Newell et al. 1992). The temporal and spatial resolution of observational data make it however impossible to infer the complex interaction between processes such as convection, rainwater evaporation, and the land surface from such an approach.

Since the advent of numerical weather prediction (NWP) models, the possibilities for studying the atmospheric water cycle have improved greatly (Andersson et al. 2005). In particular, high-resolution NWP models with advanced cloud microphysics and a realistic representation of the orography gave insight into regional-scale processes. Keil et al. (1999) used such a model to investigate the water vapour transport during the Oder flood of 1997. Zängl (2004) simulated the flood of August 2002 with a nested NWP model chain up to a resolution of 2 km. Detailed parametrisations of soil and vegetation processes allowed to study the feedbacks between soil moisture and the atmosphere (Schär et al. 1999), to explore long-term memory effects of soil moisture (Vidale et al. 2003), and to conduct sensitivity experiments involving soil moisture and vegetation (Heck et al. 2001). However, changing the model physics always implies a changed model behaviour, which can impede the straightforward interpretation of the results from such studies.

Stable water isotopes in precipitation have long been used as an independent observational method to gain insight into the atmospheric hydrological cycle (Dansgaard 1964; Rozanski et al. 1982). During evaporation and condensation processes, heavy isotopes are preferentially enriched in the liquid phase, and therefore removed from the atmosphere with precipitation. The level of isotopic depletion of the remaining water vapour can therefore provide insight into the origin and transport conditions of water vapour and precipitation. Stable water isotopes have mostly been interpreted on climate time scales, and are a key tool to interpret archives of past climate conditions, such as the sequence of glacial-interglacial cycles (Dansgaard 1993). On shorter time scales, however, the isotopic composition of precipitation can be influenced by a multitude of factors, and it is often difficult to interpret the data unequivocally (e.g. Gedzelman et al. 1989, 2003).

In order to reveal the processes that lead to observed changes in the climatic mean isotopic composition of the atmosphere, stable water isotope fractionation has been included in various atmospheric transport models. In a Lagrangian framework, isotopic fractionation was modeled first along idealised transport trajectories to the Greenland ice sheet and Antarctica (Jouzel and Merlivat 1984; Johnsen et al. 1989; Petit et al. 1991; Ciais and Jouzel 1994; Ciais et al. 1995), and recently also based on trajectories from reanalysis data for Antarctica (Helsen et al. 2004, 2005b). Stable water isotope processes have also been incorporated into several general circulation models (GCMs) (Joussaume et al. 1986; Hoffmann et al. 1998; Schmidt 1999; Noone and Simmonds 2002).

A major advantage of such isotope models is the possibility to validate the simulated hydrological cycle with observational data. On the question of precipitation origin, however, as with isotope data itself, such models remain inconclusive, as the same multitude of influences as in the atmosphere can modify the isotopic signal.

A further logical step is therefore to deploy water tracers in a GCM's hydrological cycle that carry a 'tag' (hence the name *tagging* for such methods) according to their respective evaporation source, while undergoing all hydrological processes in the model (Joussaume et al. 1986; Koster et al. 1986). While this kind of information is not available in nature, it provides the desired insight into the contributions of various moisture sources to one particular precipitation event. In an Eulerian framework, such simulations are typically conducted forward in time (Holzer et al. 2005). The Lagrangian counterpart of such a method is to identify the origin of precipitation by tracing an air parcel that is part of a precipitating air mass backward in time (Wernli 1995, 1997). Previous Eulerian and Lagrangian approaches aiming at the identification of the origin of moisture with water tracers are briefly reviewed in the following two sections.

1.1 Eulerian tagging approaches

In simple terms, the Eulerian tagging can be thought of as releasing dye into a model's hydrological cycle. The dye is redistributed with the water throughout all phase changes, while keeping the information of its origin. Tagged water vapour was first implemented in GCMs simultaneously by Joussaume et al. (1986) and Koster et al. (1986). The latter implementation was used later for a study on isotopic signals in Antarctica (Koster et al. 1992), for modeling the water sources of the Sahel (Druyan and Koster 1989), and for a study of Greenland precipitation sources under current and last glacial maximum climate conditions (Charles et al. 1994). Numaguti (1999) implemented a water vapour tagging into the CCSR/NIES GCM, and conducted studies with a focus on Eurasia. (Werner et al. 2001) conducted combined tagging and stable isotope simulations with the ECHAM-4 model to identify the moisture sources of Greenland and Antarctica. Bosilovich and Schubert (2002) implemented water vapour tagging (or Water Vapour Tracers (WVT), as introduced by these authors) in NASA's GEOS-3 GCM for a study of the North American and Indian hydrological cycle and the vertical distribution of WVT over North America (Bosilovich 2002). Bosilovich et al. (2003) implemented the same approach into the NASA FVGCM (Finite Volume GCM) to study water sources of the North American monsoon.

Currently, no regional model is known that includes such water vapour tracers. This comes as a surprise, considering the potential such a tagging RCM holds for greatly improving the insight into the regional hydrological cycle, mainly due to the higher resolution and more detailed parametrisations of sub-grid scale processes. Further useful applications of such a tagging RCM, besides the identification of moisture sources, include the assessment of processes such as precipitation recycling or soil water memory effects, as well as the investigation of model deficiencies.

1.2 Lagrangian moisture diagnostics

Information on the sources of moisture have also been gained in a Lagrangian framework. Here, air parcels are traced as they are transported in the atmosphere and, to a first order approximation, can change their water vapour content due to precipitation and evaporation processes. Wernli (1997) and later Eckhardt et al. (2004) derived quantitative precipitation estimates from back-trajectories, while leaving the question of moisture origin largely unaddressed. Massacand et al. (1998) inferred a Mediterranean moisture source for heavy precipitation on the Alpine southside from examining the specific humidity traced along back-trajectories. A first Lagrangian moisture source diagnostic was developed by Dirmeyer and Brubaker (1999). These authors used quasi-isentropic back-trajectories in combination with model-derived surface fluxes to determine evaporation sources along back-trajectories. The same method was later applied by Brubaker et al. (2001) and Reale et al. (2001), but suffered from methodological (no kinematic trajectories) and conceptual (large vertical distance between evaporation sources and air-parcel locations) shortcomings. Wernli et al. (2002) made first attempts to identify links between evaporation from an area of anomalously warm SSTs and a severe winter storm with back-trajectories. James et al. (2004) diagnosed net water change along a large number of back-trajectories to infer the moisture sources for the Elbe flood (August 2002). Later, Stohl and James (2004, 2005) applied the same methodology to a particle model, first again to study the Elbe flood, then to determine the moisture budgets of large river basins.

All current Lagrangian approaches are however limited with respect to the definite demarcation of moisture sources. A Lagrangian methodology which considers moisture changes along the full transport history of an air parcel should be able to infer quantitative information on the contribution of various moisture sources along the transport path from back-trajectories. The main advantage of such a Lagrangian moisture diagnostic is that it would parallel the information gained from Eulerian tagging methods, and hence be comparable to the results from these other approaches.

1.3 Aims and objectives of this study

The overall aim of this work is to enhance the understanding of the atmospheric hydrological cycle by identifying the sources and transport paths of tropospheric water vapour. As has been shown in the discussion above, current Lagrangian and Eulerian methods can be further improved in various aspects. At the heart of this work is therefore the development of two improved methods. This will also allow for an initial comparison of the two approaches, and give insight into their respective strengths and weaknesses. The two methods are then applied to study processes in the hydrological cycle at a global and regional scale.

More specifically,

the Lagrangian approach is extended to allow for a definite demarcation of moisture sources along back-trajectories. The methodology is applied for examining NAO variability of the moisture sources of the Greenland plateau. In order to make the results comparable to observations, the diagnosed transport conditions of moisture are then used for calculating the stable isotope composition of precipitation from a widely used Lagrangian stable isotope model. The identification of the moisture sources of the Greenland plateau can have important implications for the interpretation of stable isotope records in this area, which are key recorders of past climate change and climate variability in the Northern Hemisphere. As a side-issue, the Lagrangian method is further modified to study the conditions of Saharan dust transport to and deposition at an Alpine glacier (see Appendix A).

The Eulerian approach is for the first time implemented in a regional climate model. This requires a detailed consideration of suitable advection numerics, and other aspects of the implementation in different parts of the model. The method is then applied to study the moisture sources of the largest flood event in central Europe in recent decades, namely the Elbe flood in August 2002.

In the following chapter (Chapter 2), a review of the stable isotope processes in the atmosphere is provided, with a focus on understanding the emergence of longer-term isotope signals from a process-based point of view. In Part I, first, a Lagrangian moisture source diagnostic is developed, and applied to identify the sources of winter-time moisture transport to Greenland (Chapter 3). In Chapter 4, the identified sources and transport conditions of moisture are used to model the NAO variability of stable water isotopes in precipitation over the Greenland plateau. An extension of the Lagrangian method to study the transport of Saharan dust to an Alpine glacier is presented in Appendix A. In Part II, a new WVT algorithm is developed and tested in a 1-dimensional setting (Chapter 5), then detailed with respect to the implementation in the RCM model (Chapter 6). In Chapter 7, the developed tagging RCM is applied to simulate the Elbe flood. Finally, some concluding remarks on the main findings in this work, and a short comparison of both modelling approaches, are provided in Chapter 8.

Chapter 2

Stable Isotopes of Water

Stable water isotopes are a very powerful means to study the global water cycle, and a corner stone of paleoclimate reconstructions. This chapter reviews the concepts and principles of stable isotopes in the atmospheric water cycle. Thereby, the focus is on how synoptic variability of the water cycle establishes longer-term mean isotope signals. First, the background of stable isotope physics and some definitions will be given. The four sections thereafter describe stable isotope processes in the atmospheric branch of the hydrological cycle, advancing from the microphysical, to the synoptic, the seasonal, and finally the multi-annual scale. At the end, the current modelling efforts are briefly summarised, and implications for this work are derived. Throughout the chapter, the focus is on pointing out current knowledge gaps, and on highlighting where our trajectory-based methodology fits in.

2.1 Background and definitions

Physically, stable isotopes are atoms which take the same position in the table of elements, but have a different number of neutrons and therefore mass. The most relevant isotopes for atmospheric and hydrologic sciences are ^{18}O for oxygen (corresponding to the most abundant isotope ^{16}O), and ^{2}H (or Deuterium, D) for hydrogen (corresponding to the most abundant isotope ^{1}H) (e.g. Gat 1996; Mook 2001). The respective heavy water molecules are then $H_2^{18}O$ and HDO. These stable isotope molecules are summarised under the term *stable water isotopes*. The stable isotope ^{17}O and the radioactive isotope ^{3}H play a minor role for the present discussion, and are hence not considered further (see Mook (2001) for a detailed discussion).

The larger number of neutrons lends a larger atomic weight to the heavy isotopes. This increased mass can induce measurable physical and chemical effects. During phase changes, such as evaporation and condensation, stable isotopes become enriched in one phase and depleted in the other. This separation of isotopes between reservoirs is termed *isotopic fractionation*. Disentangling how exactly the various atmospheric processes lead to isotopic fractionation is the core problem of stable isotope meteorology.

The heavy stable water isotopes are considerably less abundant than the most common isotopes (Table 2.1). For quantifying the abundance of an isotope in a reservoir, the large difference between, for example, the number of $H_2^{16}O$ and $H_2^{18}O$ molecules in a given volume of water makes it favourable to introduce the isotopic ratio R. R is calculated as the ratio of concentrations between the rare and abundant molecules, e.g.:

$$^{18}R = \frac{\text{rare isotope abundance}}{\text{abundant isotope abundance}} = \frac{[H_2^{18}O]}{[H_2^{16}O]}. \tag{2.1}$$

Fractionation processes leading to isotopic enrichment and depletion of a reservoir can be quantified more intuitively if isotope ratios are expressed relative to a standard. Measurements of stable isotopes are also typically require a common standard. For atmospheric applications, the usual standard is the Vienna Standard Mean Ocean Water (V-SMOW), published and distributed regularly by the IAEA[1], Vienna. The delta (δ) notation is used to quantify stable isotope as relative ratios:

$$\delta^{18}O = \left(\frac{^{18}R_{\text{sample}} - ^{18}R_{\text{std}}}{^{18}R_{\text{std}}}\right) \times 1000 = \left(\frac{^{18}R_{\text{sample}}}{^{18}R_{\text{std}}} - 1\right) \times 1000 \ (‰). \tag{2.2}$$

Here R_{std} is the isotopic ratio of the standard. As atmospheric delta values are generally small, they are usually expressed in permil (‰). Typical isotopic ratios are compiled in Table 2.2.

2.1.1 Equilibrium fractionation

Isotopic fractionation in the atmosphere is to a good approximation a purely physical process, governed by the mass of the involved molecules. Isotopically heavier water molecules have two main properties which influence their fractionation behaviour: (i) due to the heavier weight, their diffusion velocity is slower, and (ii) due to their larger mass, phases with stronger bonds are preferred (solid > liquid > vapour). As the phase changes are fully reversible, given sufficient reaction time an isotopic equi-

[1] International Atomic Energy Agency

Table 2.1: Natural abundances of oxygen and hydrogen isotopes. After Mook (2001).

Oxygen		Hydrogen	
Isotope	Abundance (%)	Isotope	Abundance (%)
^{16}O	99.76	^{1}H	99.985
^{17}O	0.038	$^{2}H(D)$	0.015
^{18}O	0.200	$^{3}H(T)^*$	$< 10^{-15}$

*radioactive isotope

Table 2.2: Typical natural isotopic ratios of water in the hydrological cycle. After Mook (2001).

Natural reservoir	$\delta^{18}O$ (‰)	δD (‰)
Ocean water	-6...+3	-28...+10
Arctic sea ice	-3...+3	0...+25
Marine moisture	-15...-11	-100...-75
Lake Chad	+8...+16	15...+50
Alpine glaciers	-19...-3	-130...-90
Greenland	-39...-25	<-150...-100
Antarctica	-60...-25	<-150...-100
(Sub)Tropical precipitation	-8...-2	-50...-20
Mid-latitude rain*	-10...-3	-80...-20
Mid-latitude snow*	-20...-10	-160...-80

* from the summer/winter precipitation at the IAEA station Vienna.

librium between two phases will be reached. Consequently, these two mass-dependent fractionation effects are termed *equilibrium effects*.

As an illustration, consider the process of evaporation of water vapour from a water surface. Let the water surface have the isotopic composition of standard mean ocean water (by definition $\delta^{18}O = 0‰$ and $\delta D = 0‰$). Isotopically lighter water molecules will evaporate preferentially, and relative to the standard, form depleted atmospheric vapour, and leave behind an enriched water body. In terms of the δ notation, this is expressed as $\delta_{\text{vapour}} < 0‰$ and $\delta_{\text{surface}} > 0‰$. Vice versa, when atmospheric water vapour condensates into droplets, the heavier water molecules will preferentially form the liquid (or solid) phase, leading to enriched precipitation, and leaving behind a depleted vapour phase ($\delta_{\text{vapour}} > \delta_{\text{rain}}$).

The phase change equations for the equilibrium reactions that occur in a mixed-phase cloud, for example, can be formulated for the molecule HDO as:

$$HDO_{(v)} \leftrightarrow HDO_{(l)} \tag{2.3}$$

$$HDO_{(l)} \leftrightarrow HDO_{(s)} \tag{2.4}$$

$$HDO_{(v)} \leftrightarrow HDO_{(s)} \tag{2.5}$$

where v, l, and s denote the vapour, liquid, and solid water phases, respectively. In atmospheric conditions, the equilibrium fractionation effects depend only on temperature. Their effect is most pronounced for cold temperatures, and fades towards warmer temperatures as an effect of the increasing excitation of rotational and vibrational modes within molecules.

The stable isotope ratios of two phases in isotopic equilibrium are compared by means of the *fractionation factor* α, which effectively is a reaction constant of the equilibrium reaction:

$$\alpha_{v/l} = \frac{^{18}R_v}{^{18}R_l} = \frac{\delta_v^{18}\mathrm{O} + 1000}{\delta_l^{18}\mathrm{O} + 1000} \tag{2.6}$$

Typically, α is very close to 1.0, which is why frequently fractionation is expressed either as $1000 \times \ln(\alpha)$ or as ϵ:

$$\epsilon = (\alpha - 1) \times 1000. \tag{2.7}$$

In Eq. 2.6, $\alpha < 1$ (or $\epsilon < 0$) indicates that ^{18}O is enriched in the liquid phase; conversely, $\alpha > 1$ (or $\epsilon > 0$) would signify that ^{18}O is enriched in the vapour phase. Fractionation factors are an important ingredient for many stable isotope models. The influence of air temperature on isotopic fractionation, for instance, can easily be introduced by a temperature dependency of α. For the liquid to vapour transition of ^{18}O, for instance, Majoube (1971) determined an equilibrium fractionation factor α of 1.0098 and 1.0117 at 20° C and 0° C, respectively (see also Table 2.3). It is important here to note that due to the smaller molecule size and binding energies, fractionation factors are roughly 8 times stronger for D ($\epsilon = 0.08$ at 20°C).

2.1.2 Rayleigh distillation

A Rayleigh distillation process serves as an illustrative example for a simple model of isotopic fractionation with one reservoir and one sink (e.g. Mook 2001; Gat 1996). Consider a drop of rain that is evaporating while falling through a layer of unsaturated air. Given sufficient time, the rain drop will evaporate completely. Initially, the rain drop consists of N_0 molecules of water, and an isotope ratio of $^{18}R_0$ (^{18}R is defined as in Eq. 2.1). At a certain instant in time, the rain drop will consist of $N/(1 +^{18} R) \approx N$ molecules of the abundant and $N_i =^{18} R \cdot N/(1 +^{18} R)$ molecules of water with the rare isotope. Given the removal of dN molecules of water with the constant fractionation factor α, the mass balance for the rare isotope is

Table 2.3: Fractionation factors of ^{18}O dependent on temperature for the vapour/liquid and solid/liquid transition of water, and the ratio of ^{2}H to ^{18}O fractionation. After Majoube (1971); Mook (2001).

T (°C)	$^{18}\epsilon_{v/l}$ (‰)	$^{18}\epsilon_{l/s}$ (‰)	$^{18}\alpha_{v/l}$ (‰)	$^{18}\alpha_{l/s}$ (‰)	Ratio
0	-11.55	-34.68	0.9885	0.9653	8.7
10	-10.60	-32.14	0.9894	0.9679	8.4
20	-9.71	-29.77	0.9903	0.9702	8.1
30	-8.89	-27.56	0.9911	0.9724	7.7

$$\frac{^{18}R}{1+^{18}R}N = \frac{^{18}R + d^{18}R}{1+^{18}R + d^{18}R}(N + dN) - \frac{\alpha^{18}R}{1+\alpha^{18}R}dN. \tag{2.8}$$

Taking all denominators equal to $^{18}R + 1$ and neglecting products of differentials, this equation becomes

$$\frac{d^{18}R}{^{18}R} = \frac{dN}{N}(\alpha - 1). \tag{2.9}$$

With the initial conditions N_0 and $^{18}R_0$, Eq. 2.9 can be integrated to

$$^{18}R =^{18} R_0 \left(\frac{N}{N_0}\right)^{\alpha-1} =^{18} R_0 \cdot f^{\alpha-1} \tag{2.10}$$

or, in δ notation

$$\delta^{18}\text{O} = \frac{1+\delta_0^{18}\text{O}}{f^{\epsilon}} - 1. \tag{2.11}$$

Here, $f = N/N_0$ is the fraction of the reservoir (the rain drop) which is remaining during the Rayleigh distillation. Fig. 2.1 shows the temporal evolution of the isotopic

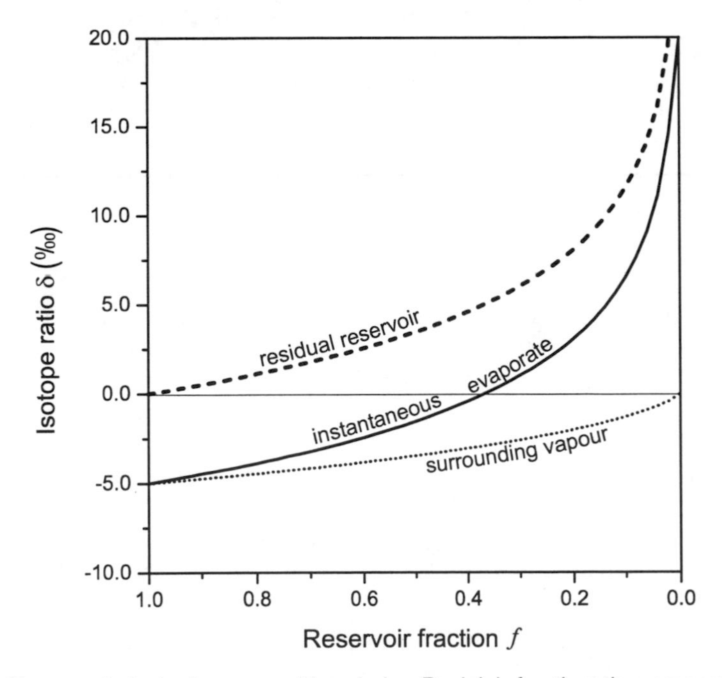

Figure 2.1: Changes in isotopic composition during Rayleigh fractionation process in a closed system. The reservoir has an initial isotope ratio $\delta_0{}^{18}$O$= 0.0$. The fractionation factor is $\alpha = 1.005$ ($\epsilon = 5$). See text for details.

composition of the rain drop and the ambient water vapour, taking the reservoir fraction f as a (non-proportional) axis of time. The rain drop has an initial isotope ratio $\delta_0{}^{18}$O= 0.0, and $\alpha = 1.005$ (or $\epsilon = 5$). The residual rain drop has an increasingly enriched isotopic composition during the fractionation process (Fig. 2.1, dashed line), as the light isotopes evaporate preferentially. The isotopic ratio of the instantaneous evaporate (Fig. 2.1, solid line) however also depends on the isotopic composition of the reservoir, and hence becomes increasingly enriched during the distillation. Note that the isotopic offset (Fig.2.1, arrow) according to the fractionation factor is constant with f. The isotopic composition of the total surrounding vapour (Fig. 2.1, dotted line) accordingly increases from the smallest value δ^{18}O= -5‰ for $f = 1$ to zero (the initial composition of the liquid phase) at $f = 0$, i.e. when the rain drop has completely evaporated.

Rayleigh distillation models are widely applied in isotope studies in the atmospheric sciences (see Section 2.6), and can be further extended by considering several coupled reservoirs, and modified to describe open systems systems (Gat 1996; Mook 2001).

2.1.3 Non-equilibrium (kinetic) fractionation

In addition to the equilibrium fractionation effects, so-called *kinetic* or *non-equilibrium fractionation* can take place during phase changes. If the evaporating water vapour above a water surface is for example continuously transported away by turbulent processes, the phase change reaction cannot reach an equilibrium, i.e. it is forced towards one side of the reaction equation, e.g.

$$\mathrm{HDO}_{(l)} \rightarrow \mathrm{HDO}_{(v)}. \tag{2.12}$$

In the transition from the saturated water surface to the turbulent boundary layer, water vapour passes through an intermediate layer where molecular diffusion velocities are important. Diffusion velocities are different for the two water isotope molecules HDO and $H_2^{18}O$. Unlike in equilibrium fractionation, there is not enough time, mostly for the slower-moving $H_2^{18}O$ molecules, to reach an equilibrium state, which results in measurable deviations from equilibrium conditions. Namely, the fractionation ratio of the two molecules will deviate from the ~1:8 ratio observed under equilibrium conditions (Table 2.3). The effects of non-equilibrium fractionation is quantified as the *Deuterium excess* (d-excess) (Dansgaard 1964). The d-excess is defined by the deviation of the combined isotopic information of δD and δ^{18}O from the relative fractionation under normal conditions:

$$d = \delta\mathrm{D} - 8 \cdot \delta^{18}\mathrm{O} \;\; (‰). \tag{2.13}$$

The extent of non-equilibrium fractionation during evaporation is influenced by various factors, such as the relative humidity gradient above the water surface, air and water temperature, and the evaporative cooling of the water surface (Merlivat and Jouzel 1979; Cappa et al. 2003).

Some controversy surrounds the interpretation of this secondary isotope parameter. While usually being interpreted as a signal of the evaporation conditions, often in particular as sea surface temperature signal (Barlow et al. 1993; Delaygue et al. 2000), kinetic fractionation does also occur during other atmospheric processes, e.g. in mixed-phase clouds during ice supersaturation (Ciais and Jouzel 1994; Ciais et al. 1995, see below).

Now that the basic notions of stable isotope processes have been laid out, a detailed and quantitative review of fractionation processes is undertaken in the following section.

2.2 Isotope fractionation processes

This section provides an overview of the physical mechanisms of stable water isotope fractionation that are relevant for atmospheric processes. The description follows approximately a natural sequence of moisture transport, beginning with evaporation from the ocean, via condensation due to cooling, the formation of rain and possibly snow, which finally reach the ground again (Fig. 2.2). Important differences between equilibrium and non-equilibrium (kinetic) fractionation effects as mentioned before will be highlighted during this discussion.

Evaporation from a water surface (Fig. 2.2, ①) comprises a combination of equilibrium and kinetic fractionation effects, following the widely used model of Craig and Gordon (1965). Directly at the interface of water and air, the two phases are in isotopic equilibrium, with the vapour being depleted by about 9‰ in δ^{18}O and 80‰ in δD with respect to the sea surface. A layer which is dominated my molecular diffusion froms the transition from the sea surface to the turbulent boundary layer. The diffusion speed in this layer is strongly influenced by the relative humidity gradient, but also by the evaporative cooling of the uppermost 0.5 mm of the water surface, as a recent study has revealed (Cappa et al. 2003). Within the turbulent surface layer, water vapour is assumedly transported without fractionation (Fig. 2.2, ②). Problems of the Craig-Gordon approach occur when additional evaporation processes come into play, in particular from sea spray and droplet evaporation, for example during strong wind conditions (Lawrence et al. 1998; Gat et al. 2003). Then, water vapour above the sea surface may be less depleted than one would expect from the Craig-Gordon model. As diffusion is important in the evaporation process, non-equilibrium fractionation takes place. The d-excess (Eq. 2.13) is hence assumed to be largely determined by the relative humidity and the SST at the evaporation site (Merlivat and Jouzel 1979; Jouzel and Koster 1996). However, the available data of water isotopes in the marine boundary layer barely allow to verify the theory (Gat et al. 2003).

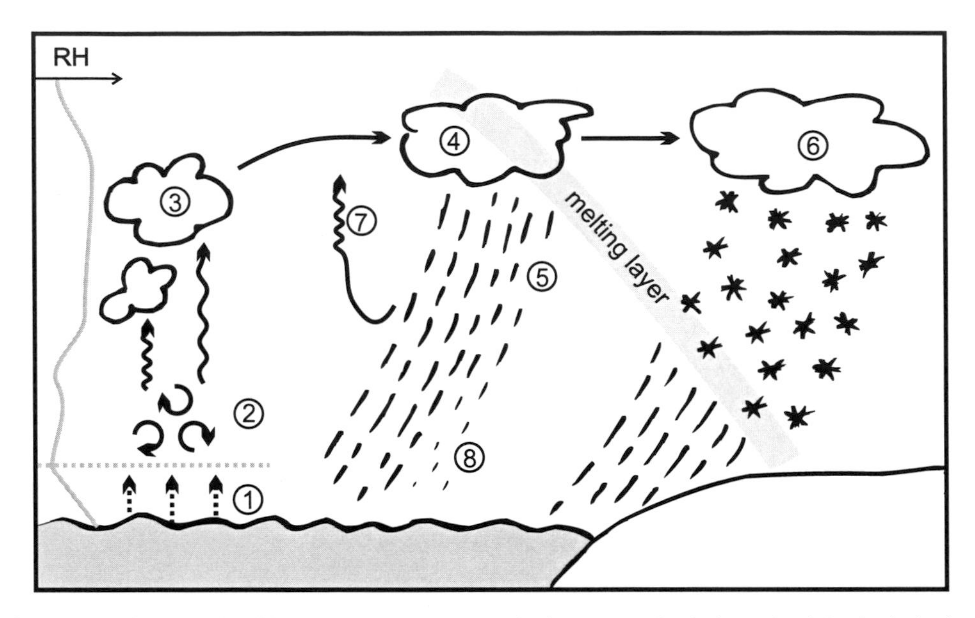

Figure 2.2: Schematic of fractionation processes in the atmospheric branch of the hydrological cycle. See text for details.

Evaporation and transpiration from land surfaces is generally assumed to occur without fractionation (Rozanski and Sonntag 1982). This assumption is justified to some degree, since, for example, intercept storage water often evaporates completely and hence the vapour retains the isotopic signature of the liquid (see Sec. 2.1.2). However, some fractionation may occur during soil water transpiration (Aggarwal et al. 2004), and little is currently known about the fractionation during the transpiration of plants. For the present study, the Ocean is by far the dominant moisture source, hence this issue is of minor importance.

As an air parcel rises, atmospheric water vapour condenses due to adiabatic and radiative cooling (Fig. 2.2, ③). The formation of cloud water is an equilibrium process, that can realistically be described by Rayleigh distillation models (Rozanski and Sonntag 1982). During continuing condensation, the heavy isotopes preferentially form the condensed phase that eventually falls as precipitation (Fig. 2.2, ⑤), thereby depleting the remaining water vapour and cloud water (Fig. 2.2, ④). The same applies when ice crystals form, yet with different fractionation factors (Fig. 2.2, ⑥). At very low temperatures (below 250 K), an additional fractionation mechanism comes into play, during diffusion of water vapour onto ice crystals. The non-equilibrium fractionation effect during this process is not well constrained by observations, and usually parameterised by a semi-empirical supersaturation function S which describes the Bergeron-Findeisen process (Ciais and Jouzel 1994).

Liquid precipitation that falls through the atmospheric column below cloud base is subject to two fractionation processes: (i) rain water equilibrates its isotopic composition with the surrounding moisture (Fig. 2.2, ⑤). The equilibration time depends on drop size and relative humidity. As the moisture at lower levels is typically less depleted than the precipitation at cloud base, the precipitation becomes enriched, and the isotopic composition of the rain at the surface closely resembles that of the moisture near the surface (Gedzelman and Arnold 1994). The larger the rain drops, the higher the fall velocity, and the lower the exchange with the surrounding moisture. Typical relaxation times and fall velocities are given in Table 2.2. (ii) When precipitation falls through layers of low relative humidity, evaporation from falling droplets leads to an enrichment of the remaining rain, and consequently significantly enriched isotopic signals at the surface (Fig. 2.2, ⑧). This can in particular be the case in the vicinity of desert areas (Gat 1996).

Already depleted water vapour can be fed back into the condensation process (Fig. 2.2, ⑦), which leads to further depletion of the air masses, and the eventually formed precipitation (Gedzelman and Arnold 1994; Lawrence et al. 1998).

Solid precipitation does not equilibrate with the surrounding moisture after the deposition on snow crystals in the cloud. Hence, snow typically communicates a significantly more depleted isotopic signal from the cloud base to the surface (Table 2.2). Hail stones may contain a sequence of information from different atmospheric levels (Jouzel et al. 1975; Federer et al. 1982b,a). Unknown isotopic effects may occur during the riming of snow crystals and partial melting. The altitude of the melting layer plays an important role for the transition between the isotopic signals of rain and snow. It determines the distance that is available for equilibration between the falling precipitation and the surrounding moisture (Gedzelman and Arnold 1994). The closer the melting layer is to the surface, the more will the precipitation signal resemble that of snow (Fig. 2.2). So far, this phenomenon has not yet been studied in the field.

The most important consequence of this discussion for high latitudes (and altitudes) is that the signal of $\delta^{18}O$ and δD in snow should in general represent the conditions at precipitation formation, while for rain it more or less reflects the conditions near the surface. As is detailed in the next section, these general characteristics are the key to understand the complex isotopic signals generated by weather systems.

Table 2.4: Relaxation times, relaxation distances, and fall velocities for typical rain drop sizes at 10°C. After (Mook 2001; Houze Jr. 1993).

Drop radius (mm)	Relaxation time (s)	Relaxation distance (m)	Terminal velocity (ms^{-1})
0.1	7.1	5.1	1.00
0.5	92.5	370	2.00
1.0	246	1600	3.50

2.3 Isotopic processes in weather systems

The fractionation processes acting on microphysical scales are usually part of well-defined weather systems. Mid-latitude cyclones, fronts, and moist convective systems are the most important organised moisture transport systems. They contain a three-dimensional water isotope information, which is to some extent communicated to the surface via precipitation processes. In this section, the principal properties of stable isotope ratios in (i) mid-latitude cyclones and their stratiform precipitation, and (ii) convective systems are synthesised.

2.3.1 Mid-latitude cyclones

Early studies of stable isotopes in mid-latitude precipitation showed a large variability between storms, and revealed in addition distinct temporal and spatial trends and fluctuations within individual storms (Dansgaard 1964; Gedzelman and Lawrence 1982; Gedzelman et al. 1989; Gedzelman and Lawrence 1990). Gedzelman and Lawrence (1990) and Gedzelman and Arnold (1994) introduced an instructive interpretation of within-storm stable isotope variability by considering the vertical structure of stable isotope composition in precipitating systems.

In a schematic 2-dimensional model[2], a cross-section through the warm and cold front of a well-developed mid-latitude cyclone separates warm and cold air masses (Fig. 2.3). At the warm front, warm and moist air is gently sliding upward onto colder and denser air-masses downstream. Large-scale condensation and fractionation occurs in this region, and water vapour increasingly becomes depleted in heavy isotopes as it is lifted to higher altitudes. A transect of isotope measurements within the clouds from downstream towards the location of the warm front at the surface would record increasingly less depleted isotopic ratios as the cloud base approaches the surface (Fig. 2.3, sector c). However, it is crucial to distinguish here between solid and liquid precipitation, or, more specifically, to consider the altitude of the melting layer. Snow will transmit the isotopic signal from the deposition altitude in the cloud to the surface without changes, while rain will tend to equilibrate with the surrounding water vapour depending on fall distance and droplet size (Table 2.2). This together with the position of the melting layer and the decreasing depletion of the cloud water leads to the gradual increase of the isotopic signal in surface precipitation towards the surface warm front (Fig. 2.3, right curve in lower panel). Dansgaard (1964) already noted the height of the front above ground as an important parameter for the stable isotope composition of precipitation, but the height of the melting layer should be considered as an equally important parameter (Gedzelman and Arnold 1994).

[2]Note that this view of a mid-latitude cyclone is highly idealised, and does not represent the full 3-dimensional flow structure within developing cyclones (compare e.g. Wernli 1995).

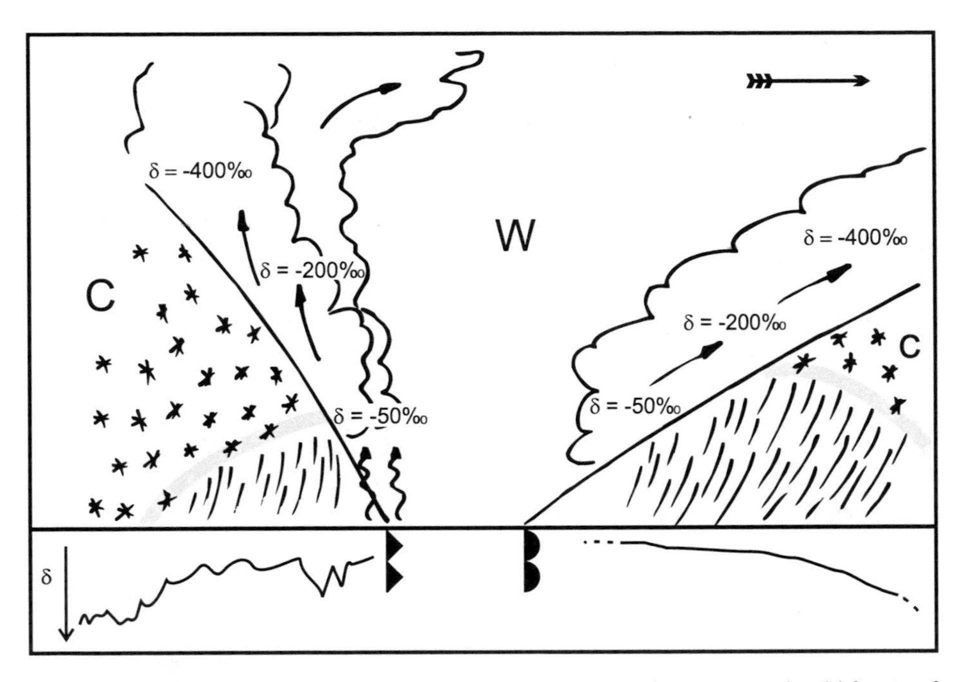

Figure 2.3: Schematic sketch of stable isotope processes at the warm and cold fronts of an extratropical cyclone. W: warm sector, C,c: cold sectors. Gray band: melting layer. δ gives typical values for δD. Lower panel sketches a typical instantaneous isotopic signature for the precipitation at the surface. Barbed arrow indicates wind direction.

Similar processes occur near the cold front of a precipitating cyclone (Fig. 2.3, sector C). In an upstream transect through the cold sector C one would observe an inverted sequence, starting with a variable isotope signal, then going from only weakly towards increasingly depleted isotope ratios. Convection at the cold front can lead to short episodes of very depleted precipitation, as is explained in the next section. In this example, the decreasing height of the melting layer in the cold sector reduces the equilibration distance for the falling rain, and then leads to significantly lower isotopic values as snow reaches the surface (Fig. 2.3, left curve in lower panel). Precipitation from occluded cyclones or with embedded convection is likely to have an even more complicated isotopic signature.

During their life-cycles, mid-latitude cyclones will produce increasingly depleted precipitation as they travel polewards along the storm tracks and transport moisture to higher latitudes. From the above discussion, it can be derived that this is due to (i) fractionation at colder air/cloud temperatures due to adiabatic and radiative cooling, (ii) more depleted water vapour due to lifted fronts, and (iii) the melting layer being closer to the ground with increasing latitudes. The role of the initial isotopic composition of the moisture is discussed in Sec. 2.3.3.

2.3.2 Convective systems

The isotopic processes in convective systems are more complex and less understood than for systems with stratiform precipitation. Conceptually, convection is a process acting mainly on the vertical scale (Fig. 2.4). Moist convection is triggered either by adiabatic, diabatic, or orographic processes in (conditionally) unstable air. Rapid vertical upward motion (updraughts, Fig. 2.4, upward arrows) leads to rapid condensation and precipitation formation with large droplets and intense showers. Precipitation falling back through the updraughts equilibrates with the surrounding vapour, and is hence in general less depleted than stratiform precipitation. Large drop sizes and high precipitation intensity however counteract equilibration. Rapid transport in cold downdraughts (Fig. 2.4, downward arrows) can in addition reduce equilibration and lead to episodes of more depleted precipitation. In the same way as depleted moisture from higher altitudes can be recycled into the convective cloud via entrainment, new depleted layers of water vapour form in the high-altitude outflow regions of convective systems (Gedzelman and Arnold 1994; Gedzelman and Lawrence 1982). Hail stones moving up and down between melting and freezing layers carry isotopic records of their formation

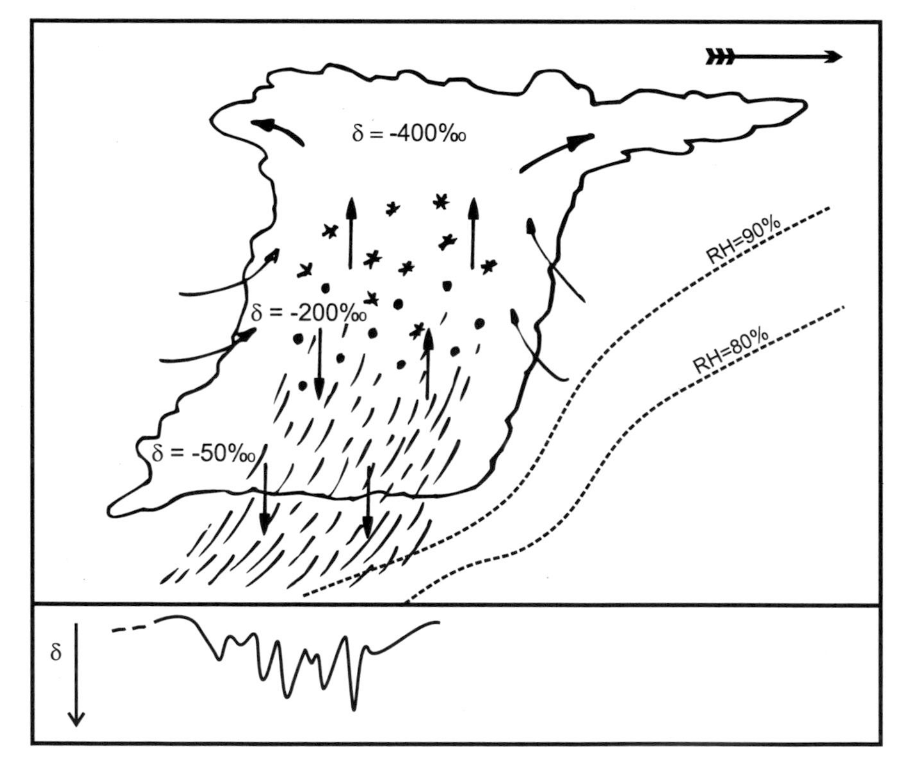

Figure 2.4: Schematic sketch of isotopes in a convective cloud. Lower panel sketches a typical isotopic signature for the precipitation at the surface. Barbed arrow indicates wind direction.

conditions to the surface (Federer et al. 1982b). If the convective cell precipitates into unsaturated air masses, rain water evaporation may again increase the isotope ratios of the remaining rain. In an upstream transect below the convective cloud (Fig. 2.4, lower panel), an observer would generally measure a very variable isotopic signal, at first with slightly enriched isotope ratios, then strong excursions towards highly depleted values in downdraught and intense precipitation regions, and finally again a less depleted tailing-out of the isotope record, as equilibration in lighter rainfall sets in again.

In the inter-tropical convergence zone (ITCZ), large-scale convective systems are responsible for by far the largest part of the precipitation. In these areas, the isotope characteristics of convective systems also dominate the stable isotope seasonality in precipitation. The moisture transport characteristics of Monsoon systems are at present not well understood. Again, such systems lead to sustained heavy precipitation with large droplets and relative humidity close to saturation below the cloud, which can allow for the communication of depleted isotopic signals from high altitudes to ground level (see amount effect, pg. 24). In large-scale basins, such as the Amazon or central China, an important fraction of precipitation is recycled from land sources, which in addition leads to successively depleted isotopes (see continentality effect, pg. 24). Tropical cyclonic systems, such as hurricanes, have been observed to produce very depleted precipitation (Lawrence and Gedzelman 1996). Besides the very effective precipitation generation of such systems, which include embedded convective cells, moisture recycling between successive rain bands centered around the wall region has been proposed as a mechanism to produce such depleted isotopic signals (Lawrence et al. 1998).

2.3.3 Further remarks

In the above discussion, the initial isotopic composition of water vapour has not been taken into consideration as a factor of influence. This influence is not well constrained by observations, and is often assumed as equal to the isotopic ratios of precipitation near the surface. The isotopic composition of δD and $\delta^{18}O$ near the surface can be modelled, for example in a general circulation model (GCM) by means of the Craig-Gordon approach (see Sec. 2.6).

All of the above processes pose particular problems for the d-excess parameter, as very little is known on non-equilibrium fractionation in clouds, and experimental data from the atmosphere are hardly available. While hence the short-term variability for this parameter in precipitation or water vapour may be impossible to interpret, on longer time scales it may still contain useful information, as will be further elucidated below.

For a Lagrangian modelling approach the implications of the above discussion of stable isotopes in weather systems are that after deposition in the cloud the isotopic ratios of snow remain unchanged and should be successfully predictable by Rayleigh models. For rain, however, the processes acting below the cloud exert an important influence on the isotopic ratios in precipitation at the surface, and should not be neglected. The preferred targets for Lagrangian isotope models are hence high latitudes, winter seasons, and high altitude locations.

In the following section, the understanding of the stable isotope signature of individual weather systems presented above will be applied to describe the mean global water isotope characteristics.

2.4 Isotopic processes on the seasonal to annual scale

On seasonal to annual time scales, a climatological mean state emerges from the superposition of weather-induced isotope signals. This isotopic mean state, in particular for precipitation and the ocean surface, reflects some important characteristics of the global hydrological cycle. Dansgaard's (1964) work on the longer-term isotopic composition in surface waters and precipitation was a corner stone to gain a semi-empirical understanding of the atmospheric branch of the hydrological cycle. In this section, this now 'classical' empirical view is re-interpreted from the viewpoint of a process-oriented understanding of stable isotope fractionation. By regarding the so-called *isotope effects* in their genetic context a physically consistent interpretation is achieved and conceptual caveats become evident.

2.4.1 Annual mean isotopic signature

Since 1961, the IAEA compiles a database containing measurements of stable isotope composition in precipitation with monthly resolution from stations distributed worldwide. This Global Network of Isotopes in Precipitation (GNIP) database[3] provides the basis for global maps of water stable isotope ratios in surface precipitation. The global maps of $\delta^{18}O$ (and δD, not shown) in precipitation exhibit several salient features (Fig. 2.5). A strong latitudinal gradient is evident, with increasingly depleted ratios towards the poles. Tropical areas show the least depleted precipitation, close to V-SMOW. Mountain ranges, in particular the Rocky Mountains and the Himalaya, form depleted regions. Finally, isotope ratios are more depleted towards the interior of the continents, for example in Eurasia. Lowest ratios are observed near the poles, with $\delta^{18}O < -24‰$ and $\delta D < -150‰$ in northern Greenland, while in central East Africa isotope ratios are close to V-SMOW. It should however be highlighted that many stations in the GNIP database only have a discontinuous record, and that due to the higher density of observations significant biases exist towards the Northern Hemisphere and, in particular, Central Europe.

[3]GNIP database accessible at the URL http://isohis.iaea.org/GNIP.asp

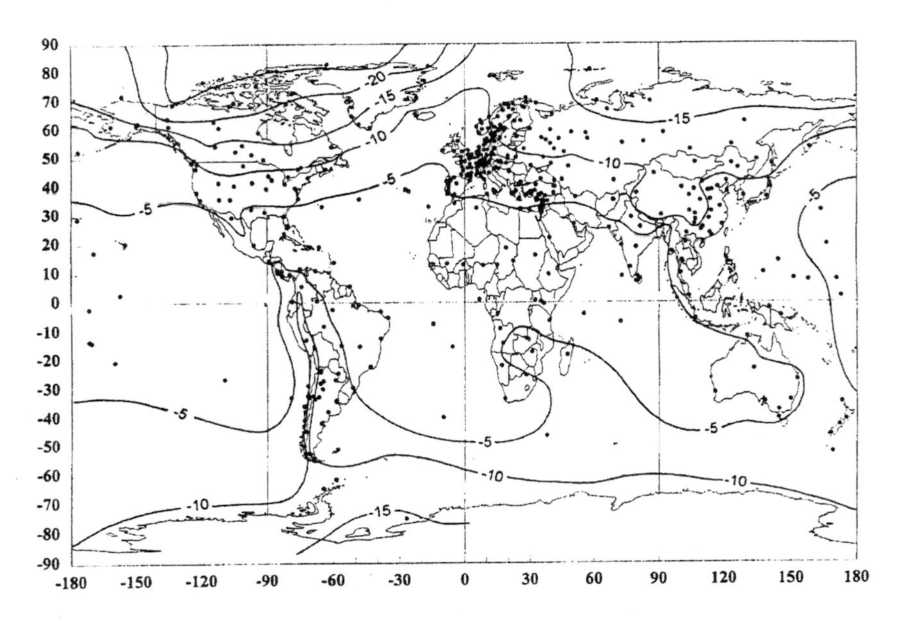

Figure 2.5: Annual mean composition of δ^{18}O in precipitation compiled from the GNIP database (Araguás-Araguás et al. 2000)

Seasonal variation is significant at some observation sites, while others have rather small annual amplitudes (Fig. 2.6). Strongest amplitudes range up to 16‰ for δ^{18}O, for example at the continental station Wynyard, Canada (51.46°N; 104.10°W). Coastal or island stations, such as Valencia, Spain (39.29°N; 0.24°W) or Midway, Pacific (28.12°N; 177.24°W), show small annual amplitudes with slightly depleted isotope ratios. A correlation between long-term mean annual averages for δ^{18}O and δD from GNIP stations results in the so-called Global Meteoric Water Line (GMWL). This regression line has a slope of ~8 and an offset of ~10‰. The GMWL underlines that in the mean, δ^{18}O and δD generally are subject to the same processes in the atmosphere, yet δD fractionation is 8 times stronger. The regression line has a 10‰ offset to the origin, which is due to kinetic fractionation effects, and hence equal to the *d*-excess. This indicates that on longer time scales the δD and δ^{18}O show persistent fractionation differences, probably due to the prevailing evaporation conditions. The mean seasonal cycle of stable isotope composition at a station generally varies along the GMWL. For some stations, significant deviations from the GMWL have been noticed, and so called Local Meteoric Water Lines (LMWL) have been established. However, deviations from the GMWL are mainly due to local processes, such as rainwater evaporation and seasonal transport changes rather than different fractionation processes, hence using LMWL can be a misleading concept (Gat 1996).

Even though by definition the mean isotopic composition of ocean water is $\delta(D,^{18}O)_{V\text{-}SMOW} = 0‰$, due to evaporative enrichment or freshwater input, individual observations of ocean surface waters can show isotope ratios ranging between -6 to

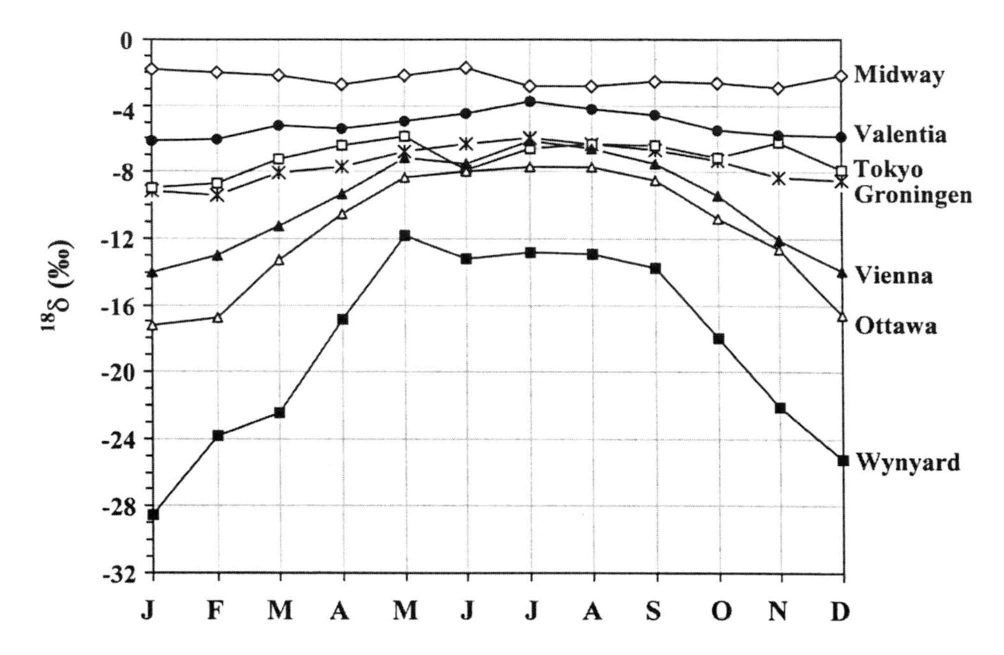

Figure 2.6: Selected seasonal cycles of δ^{18}O from Northern Hemisphere GNIP stations (Mook 2001).

+3‰ (Schmidt 1999). A map of the global mean distribution of δ^{18}O in ocean surface waters compiled from all available data (uncorrected for season) shows more depleted ratios towards the poles due to meltwater input from glaciers (Fig. 2.7). In some areas, such as the Mediterranean, net evaporation results in enriched isotope ratios of up to +3‰. Less data are available for δD; here, measurements in ocean surface waters range between -28 and $+10$‰. In general, for oceans the observational database of isotopic composition is very limited, and data gaps can only be filled by isotope GCM models (see Sec. 2.6).

2.4.2 Stable isotope effects

The global and seasonal mean picture of stable isotopes in precipitation has led Dansgaard (1964) to formulate the empirical *isotope effects*, and due to their intuitive instructiveness they have become a widely used means for interpreting stable isotopes. However, on shorter time scales, variability is considerable, the simplified mean picture disintegrates, and individual 'effects' are difficult to separate. Without a detailed process-oriented understanding, interpretation of stable isotope ratios by means of the isotope effects may even be misleading. In addition, a number of assumptions need to be made, such as a fixed moisture source for global precipitation between 40° N and 40° S, which now have partly been shown to be too simplified. Transport influences

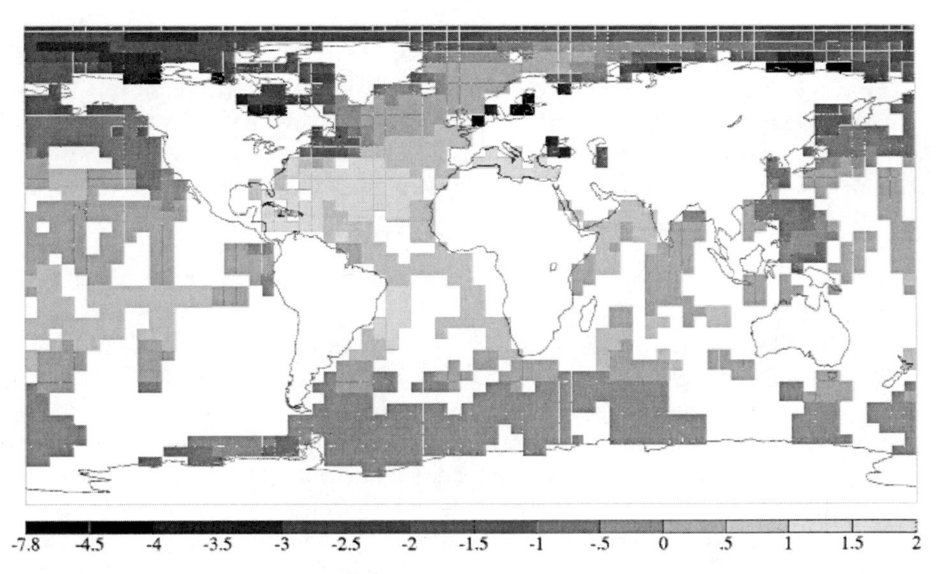

Figure 2.7: Ocean surface mean isotopic composition compiled from all available observations (Schmidt 1999).

are mostly neglected, even though they are very relevant for the interpretation of the global water isotope pattern (Rozanski et al. 1982).

In the following, Dansgaard's isotope effects are listed and discussed from a process-oriented perspective.

Latitude effect: The latitude effect is based on the observation of a gradually depleted isotope signal in precipitation towards higher latitudes (Fig. 2.5). The average gradient for δ^{18}O is $-0.6‰/°$ latitude in mid-latitudes, and up to $-2‰/°$ latitude in Antarctica (Mook 2001). Along with latitude, isotope ratios correlate with annual mean surface temperature according to the *isotope-temperature relationship* (δ-T relationship) (Eq. 2.14, Dansgaard 1964):

$$\delta^{18}\mathrm{O} = 0.62 \cdot \mathrm{T} - 15.25 \ (‰). \tag{2.14}$$

This relationship between stable isotopes and surface temperature has become one of the principal tools of paleoclimatology, primarily in the interpretation of high-latitude ice cores. In light of the previous sections, it is evident that a number of processes contribute to this latitude effect. Condensation temperature and hence fractionation factors change as moisture condensates at more northerly latitudes (Table 2.3). Precipitation increasingly falls as snow at high latitudes, and hence equilibrates less with surface moisture. Ocean surface water is more depleted in high latitudes, and so will be the evaporating moisture from it. Cyclonic systems from the mid-latitudes produce increasingly depleted precipitation as they travel northward along the storm track. All of these factors contribute to more depleted isotopic ratios. Considering that the simple δ-T relationship emerges from the interplay of a number of processes which cannot *a priori* be regarded as time-invariant on climate time scales (Brown and Simmonds 2004), which underlines the importance of process-oriented investigations of isotope fractionation.

Altitude effect: The isotopic ratios of precipitation generally decrease with increasing altitude (Fig. 2.5, e.g. the Andes). The decrease ranges between -0.1 and -0.6‰/100 m, in the western U.S. and Chile it can be up to -4‰/100 m (Mook 2001). The reasons for this altitude effect are again complex. It can be directly related to the decrease in saturation vapour pressure with altitude, i.e. when precipitation forms orographically and condensation takes place at different vertical levels. In addition, the temperature dependence of fractionation factors comes into play. However, the equilibration length of falling rain drops is different for different altitudes as well, and the height of the melting layer can be of significant influence in highly orographic terrain. Finally, different transport patterns may govern moisture advection at different altitudes, for example when precipitation occurs below or above the trade inversion, or has a different seasonality with altitude.

Continental effect: In general, isotopic ratios decrease with increasing distance from the coast (Fig. 2.5, e.g. in Eurasia). Prominent examples are transects from Valencia to the Ural mountains, which show a decrease of -12‰ in $\delta^{18}O$ over a distance of 5000 km (Rozanski et al. 1993). Again, multiple factors contribute to this continentality effect. As a precipitating system moves inland, the precipitation becomes increasingly depleted as does the remaining water vapour. In addition, depleted terrestrial moisture may be recycled and further contribute to depleted isotope ratios. The annual temperature amplitude increases with increasing continentality, which influences the seasonality of condensation temperatures and hence fractionation factors. The contribution of convective systems to precipitation may increase further inland. Finally, the seasonality of precipitation may be different, and, what leads to a further depleted isotopic signal, a larger fraction of solid precipitation may be deposited during wintertime at continental sites. The continental effect is hence strongly interlinked with the seasonal effect (see below).

Amount effect: Stations with high annual precipitation amounts can on some occasions exhibit significantly depleted isotopic ratios (Fig. 2.5, e.g. in Indonesia). This amount effect is in particular found in areas affected by the ITCZ, as well as in hurricanes and, to a lesser extent, in other convective precipitation regimes. The highly depleted values in heavy precipitation are physically related to low rainwater-equilibration due to large droplets and high relative humidity below convective clouds, downdraughts that decrease the time available for equilibration, and finally the recycling of precipitation, e.g. in the Amazon basin, and as hypothesised by Lawrence et al. (1998) also between successive rain bands (see Sec. 2.3.2).

Seasonal effect: The seasonal variation of stable isotopes in precipitation apparent at many observation sites is termed seasonal effect (Fig. 2.6). Different stations show largely different seasonalities, for example with annual amplitudes of up to 16‰ in Wynyard, Canada. Generally interpreted as a temperature-related effect on the fractionation factors, other influences are often also relevant. The precipitation type can be very different between seasons, such as predominantly stratiform precipitation in winter versus convective precipitation in summer, different extent of water recycling, different melting layer altitudes, and a different ratio of solid to liquid precipitation. The seasonality of the stable water isotopes is very useful for the dating of ice core records by annual layer counting (e.g. Eichler et al. 2000; Sodemann et al. 2006). Physically, however, the seasonal effect is impossible to attribute to one single meteorological cause.

The above reconsideration of Dansgaard's classical isotope effects from a process-oriented perspective illustrates the complexity inherent to the interpretation of stable isotopes in precipitation. Furthermore, it evidences the limitations of interpreting trends in precipitation as the superposition of 'effects'. From a modelling perspective, studying the isotopic composition at cloud condensation level instead of precipitation at the surface could considerably reduce the variability, and provide a useful data set for model validation. The value of a detailed process understanding is also evident on inter-annual to climate time scales, as is elucidated in the following section.

2.5 Isotopic processes on the inter-annual to climate scale

One of the most important applications of stable isotopes in the water cycle is their use for the reconstruction of past climate variability, on inter-annual to multi-millenial time scales. The signal of $\delta^{18}O$ and δD in ice cores from Greenland and Antarctica has been used as a proxy for the polar mean temperature over the last 123'000 and 740'000 years, respectively, with seasonal to multi-year resolution (Dansgaard 1993; EPICA community members 2004). In addition, *d*-excess is usually interpreted as a moisture source region signal, primarily due to assumed influences of the relative humidity and sea surface temperature (SST) on evaporation (Johnsen et al. 1989; Barlow et al. 1993). As previously noted, observational constraints on this parameter from atmospheric measurements are however very limited.

In the absence of detailed process-based knowledge, the primary tools for the paleoclimatic interpretation of stable isotope records are the empirical isotope-temperature relationship, and the isotope effects proposed by Dansgaard (1964). However, considerable uncertainty exists on how constant this empirical relation has been through time, and on additional parameters that could influence this dependency (Jouzel et al. 1997; Brown and Simmonds 2004). Isotope GCM studies have more recently increased the understanding of isotope processes on climate time scales (see Sec. 2.6).

The main points of uncertainty are linked to the following questions:

(i) How do individual precipitation events translate into a mean value (Helsen et al. 2005a)?

(ii) What is the relation between the surface and cloud temperatures?

(iii) What is the influence of transport variability?

(iv) Were the patterns and their seasonality of atmospheric moisture transport constant on climatic time scales (e.g., Krinner et al. 1997; Krinner and Werner 2003)?

(v) Which transformations can occur in the archive after deposition (e.g., Delmotte et al. 2000)?

While questions (i)–(iii) are in the focus of the Lagrangian methodology introduced below, (iv) and (v) cannot be considered here.

Process-aimed studies are virtually impossible to accomplish on climate time scales, as direct observations are replaced by ever sparser proxy data from various archives as one goes further backward in time. One possibility to circumvent this difficulty is to use inter-annual variability of the present-day climate as a proxy for the atmospheric reorganisations that accompanied past climate variability. On inter-annual time scales, isotope data in precipitation during the last ~50 years, mainly from firn cores, are assisted by reliable observations of the general circulation, such as reanalysis data.

The North Atlantic Oscillation (NAO) is an important (mostly) atmospheric mode of variability in the Northern Hemisphere (Walker and Bliss 1928; Hurrell et al. 2003). It is profoundly linked to the general circulation, and in particular the patterns of temperature and precipitation in areas adjacent to the North Atlantic. Historically, the NAO index is defined as the difference in sea level pressure between Lisbon and Iceland, normalised by the mean state; more recent studies use empirical orthogonal functions (EOF) to extract the main modes of variability (Hurrell 1995). The NAO index oscillates between positive and negative phases, and the pattern is particularly pronounced during winter months. During summer, the NAO is not the dominant mode of variability. The mean sea level pressure pattern associated with positive NAO phases is a strong Icelandic low and a strong Azores high, while during the NAO negative phase a weak surface pressure gradient prevails over the North Atlantic (Figs. 2.8,2.9).

The NAO variability is particularly important for observations from Greenland. It considerably changes winter temperatures and precipitation around and above the

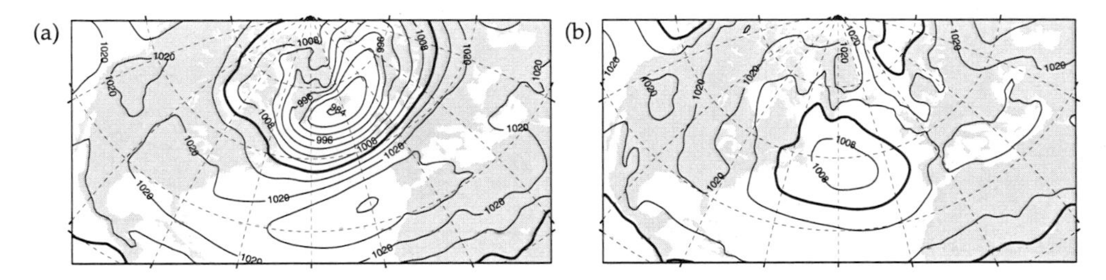

Figure 2.8: Mean sea level pressure for representative winter months for (a) positive NAO, (b) negative NAO months. See Chapter 3 for the data sources of this figure.

Greenland ice sheet (Rogers et al. 1998; Hurrell et al. 2003). During positive NAO phases, temperatures above Greenland are up to 4 K colder at a 700 hPa surface than during negative NAO months (Fig 2.9). Greenland precipitation is generally more widespread and frequent during negative NAO phases (Bromwich et al. 1999). The two main phases of the NAO have been active with varying intensity over the last 150 years covered by observations, but extend backwards considerably further in time (Vinther et al. 2003).

The inter-annual variability in the general circulation associated with different NAO phases shares similarities with assumed changes during longer-term climate shifts: Apart from the profound differences in circulation, temperatures, and precipitation, sea surface temperature can vary, adding an important factor of influence on longer time scales. Therefore, we use NAO variability here to examine pronounced atmospheric reorganisations and their impact on the stable isotope record in Greenland. It can be speculated that similar mechanisms of atmospheric reorganisation could also have been active during rapid shifts in the longer-term climate, such as Dansgaard-Oeschger events, or the Younger Dryas period during the Holocene, and may have contributed to the corresponding isotopic signals preserved in Greenland ice cores (Dansgaard 1993; Stuiver and Grootes 2000).

The NAO-induced variability of circulation, temperature, and precipitation in Greenland is also apparent in the annual snow accumulation sequence of ice cores (Appenzeller et al. 1998b,a; Mosley-Thompson et al. 2005), as well as to some extent in the stable isotope signals in firn and ice cores from central Greenland (Barlow et al. 1993, 1997; White et al. 1997). However, the NAO signal is often blurred by other influences on the stable isotopes, which requires typically to use stacked (composite) ice cores and EOF analysis to extract significant correlations with the NAO. Despite the strong influence on the circulation in Greenland, the NAO only explains about half the variance observed in stable isotope records (Rogers et al. 1998; Vinther et al. 2003). Nevertheless, accumulation sequences and stable isotope records have been used to reconstruct the NAO time series several centuries beyond the observational period.

A considerable challenge for the validity of the isotope-temperature relationship in Greenland is the discrepancy of isotope-based temperature reconstructions with tem-

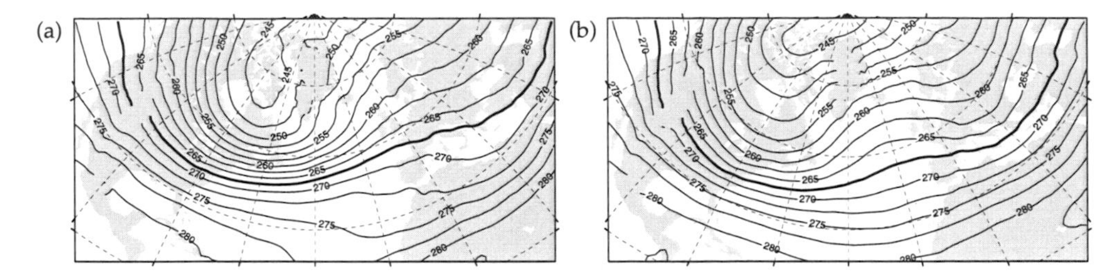

Figure 2.9: Mean temperature at the 700 hPa surface for representative winter months for (a) positive NAO, (b) negative NAO months. See Chapter 3 for the data sources of this figure.

perature estimates from bore holes (Dahl-Jensen et al. 1998; Krinner and Werner 2003). Stable isotope temperature estimates are several degrees colder during the last glacial maximum (LGM) than bore-hole temperatures. A number of hypotheses have been put forward to reconcile this discrepancy, such as changes in the position of the storm track and the polar front, a different ice sheet topography which enforced changes in the general circulation, and seasonality shifts of the precipitation maximum (Steig et al. 1994; Krinner et al. 1997; Hoffmann et al. 2000; Krinner and Werner 2003). All of these hypotheses imply profound changes in the isotope-temperature relationship. If the influence of the NAO variability on the isotope-temperature relationship can be resolved from a process-oriented perspective, it may be possible to extrapolate this knowledge to the variability on glacial-interglacial climate time scales as well. This underlines how crucial the emergence of a process-based understanding of isotope records may be for the climatic interpretation of ice core data.

The secondary parameter d-excess also shows seasonal and inter-annual variability in Greenland firn and ice-core records. This variability has been interpreted as a temperature signal of the water vapour source, assuming a constant source region of water for precipitation (Johnsen et al. 1989). With today's knowledge of atmospheric moisture transport, this concept is over-simplified, and it has been realised that transport must exert a strong influence on the d-excess (Masson-Delmotte et al. 2005a,b). How exactly the change in d-excess comes about is however still unresolved.

Processes associated with the preservation of the stable isotope signal in archives are an important source of uncertainty. A multitude of influences during and after deposition can alter the isotopic signal preserved in falling snow. Snow-drift, sublimation, sastrugi formation, firn convection, melting, flow deformation and diffusion are various issues to consider. They generally limit the temporal resolution that can be achieved with these archives to between several months to years as one goes back further in time (Seimon 2003; Helsen et al. 2005b).

An evaluation of the isotope-temperature relationship which focuses on the processes acting on an inter-annual time scale could help to clarify certain disputed aspects: What is the spatial variability of isotopic signals on the Greenland plateau? How strong is the event-to-event variability in isotopic parameters? Does the isotope-temperature relationship vary with changes in the NAO? Stable isotope models are invaluable means to accomplish this task. Their past and present use, as well as a proposed extension are compiled in the following section.

2.6 Modelling of stable water isotopes

Aiming to enhance the understanding of stable water isotope processes on local to global scales, current stable isotope models attempt to synthesise the available knowledge on stable isotope processes to various degrees. From the earliest Rayleigh-distillation models, research has advanced to GCMs which have stable isotope processes coupled into the model's hydrological cycle (Joussaume et al. 1984). The (zero-dimensional) Rayleigh model has been presented as an example in Section 2.1. In this section, details of more advanced modelling attempts are given, together with a discussion of the limitations of each model type. Furthermore, it is discussed how they can be combined to study unresolved aspects of stable water isotopes in the atmosphere.

A fundamental distinction in stable isotope modelling can be made between two model families, namely Lagrangian and Eulerian approaches. The Lagrangian category considers the fractionation of the moisture in an air parcel during its transport through space and time. Eulerian model approaches calculate isotopic fractionation based on first principles for the whole temporally and spatially discretised model domain. Both approaches are seconded by parameterisations of sub-grid scale processes. While in both categories increasingly complex models have been developed, several intermediate approaches have been proposed recently which try to combine the advantages of the two families. A description of the characteristics, strengths and weaknesses for the various existing approaches provides the framework for the approach applied in this work.

Lagrangian Rayleigh models: Lagrangian Rayleigh distillation models are commonly used in conjunction with prescribed idealised or 'climatological' pseudo-trajectories, which follow the hypothetical path of atmospheric moisture transport (Dansgaard 1964; Johnsen et al. 1989). Jouzel and Merlivat (1984) and Ciais and Jouzel (1994) included microphysics parameterisations into such a model (Mixed Cloud Isotope Model, MCIM), which allowed an application to cold polar regimes. The original idea of the MCIM model was to understand stable isotope fractionation on climate time scales, but in principle it can be applied to shorter time scales as well. The intriguing intuitiveness of this concept is alleviated by a number of difficulties. First, idealised trajectories of moisture transport do generally fall short of representing the large variability of atmospheric moisture transport, as they assume a fixed water source in the tropics. Second, due to lack of better information at that time, several parameterisations based on semi-empirical knowledge are build into the model. This concerns the relation between surface and cloud temperature and the kinetic fractionation effect in supersaturated ice clouds. Finally, the initial isotope ratios of air parcels are not well-constrained, while being quite important for the final isotope ratio (Jouzel and Koster 1996). Further details on the MCIM model are given in Appendix C.

2D column models: Two-dimensional column models have been developed mainly to increase the understanding of tropospheric isotope fractionation in the vertical (Rozanski and Sonntag 1982; Gedzelman and Arnold 1994). These models have been applied to precipitation events on time scales of hours to days, and partly contain advanced microphysics parameterisations, such as cloud ice, liquid, graupel, hail, and vapour species. The strength of these models is to increase the understanding of the vertical processes in idealised situations as they capture all processes that influence the isotopic ratio of surface precipitation. Furthermore, they help to understand observed time series of isotopes in precipitation on time scales of minutes to hours. Major limitations are the difficulty to initialise such models (e.g. from soundings), the sparsity of vertical stable isotope profiles in the troposphere to compare with, and the exclusion of isotope advection and fractionation during surface evaporation.

2D advection models: Two-dimensional isotope advection models are a recent approach to modelling stable isotopes in mid- and low latitudes. The model of Yoshimura et al. (2003) combines a moisture budget equation on a single-level $2.5° \times 2.5°$ global grid with an upstream advection scheme and Rayleigh distillation equations. Driven by reanalysis data, the model is able to reproduce surface precipitation variability to a fair degree, yet with considerable offsets. Isotope advection models highlight the importance of advection for isotopes in precipitation in temperate and tropical regions, and provide a possibility to evaluate some aspects of reanalysis data for these regions (Yoshimura et al. 2004). However, rather crude assumptions regarding important processes such as evaporation and cloud microphysics probably lead to the offset noted above, and considerably limit new process understanding by this kind of model. Kavanaugh and Cuffey (2003) developed an intermediate complexity model (ICM) that simulates isotopic fractionation along a meridional transport path, and combines the influences of advection and eddy diffusivity. From the required model tuning, they derived that diffusive 'isotopic recharge' during transport is crucial to achieve realistic results with such an ICM.

3D GCMs and RCMs: Global circulation models fitted with water isotope processes were first introduced by Joussaume et al. (1984). By now, several GCMs and one regional climate model (RCM) (Sturm et al. 2005; Fischer and Sturm 2006) are capable of simulating the stable isotope composition of the hydrological cycle in an Eulerian framework (Hoffmann et al. 1998; Schmidt 1999; Noone and Simmonds 2002). All sub-grid scale fractionation processes have to be parameterised in these models, which introduces uncertainty where process understanding is limited. This is in particular the case for mixed-phase clouds and low-temperature environments. Nevertheless, the stable isotope patterns in precipitation are well-reproduced, with some exceptions for high latitudes, and the Deuterium excess (Hoffmann et al. 2000; Delaygue et al. 2000; Werner et al. 2001). Such models can also be used to fill data-sparse regions. The process understanding gained from 3D models is despite their success limited, since the simulated isotope pattern represents the influences of different fractionation processes, and errors may mutually compensate. Due to lack of suitable observational data, validation of the simulated vertical structure is virtually impossible. Required spin-up times and low resolutions are further difficulties, in particular in polar regions, where interest is largest. Biases due to specific model climate characteristics exist, but could be circumvented by nudging GCMs with reanalysis data (G. Hoffmann, pers. comm., 2005).

Intermediate approaches: Modelling efforts which make use of both Lagrangian and Eulerian information to gain a better process understanding are considered here as intermediate approaches. Three-dimensional kinematic backward trajectories calculated from reanalysis data provide moisture transport information which can be compared with surface observations from 1957 onwards. They also capture a considerable part of the variability present on synoptic time scales. As proposed by Jouzel and Koster (1996), GCM-produced fields of stable isotope ratios of water vapour can provide initial conditions for Lagrangian fractionation studies. Helsen et al. (2004) used such an approach to study the isotope signal of a single firn core location in Antarctica. Large numbers of backward trajectories can be used to describe the fractionation history of a complete air mass. Thereby, fractionation-relevant parameters are accessible with high spatial and temporal resolution, and a degree of similarity to the actual evolution of atmospheric flow that is not available from current GCMs. Combined with a suitable moisture transport diagnostics, backward trajectories can in addition provide information on the sources and fractionation history of moisture in an air parcel. Due to the available reanalysis data, such studies are currently limited to the last 44 years, and computational constraints make such studies only feasible on monthly to multi-annual time scales.

A summary of the advantages and limitation of the different types of stable isotope models is compiled in Tab. 2.5. From the comparison of the strengths and weaknesses in current models, four main themes in enhancing the process understanding due to future stable isotope model development can be designated. These challenges are

(i) to identify the reasons for poorly reproduced patterns of non-equilibrium fractionation processes, in particular during evaporation, and to find better parameterisations for these,

(ii) to compare models with observational surface data on time scales from minutes to days,

(iii) to reveal what the importance of large-scale transport and mixing are for stable isotopes in precipitation,

(iv) to find means to compare the vertical structure of stable isotopes in the atmosphere with observational data.

The intermediate approach which is put forward here has a great potential to provide, in combination with the other modelling approaches, more process understanding along these four themes. Its main advantages are that it can be compared to current high-quality observational data, and that source-receptor relationships make an attribution of cause and effect in a Lagrangian framework very intuitive. The results drawn from an advanced Lagrangian methodology have potentially important implications for the sensitivity and limitations of Lagrangian studies using climatological mean wind and temperature information. In the following two chapters, it is demonstrated that from a study of inter-annual variability in the Northern Hemisphere with a focus on moisture transport to the Greenland ice sheet, such an intermediate approach does indeed yield new valuable insights into the climatological interpretation of stable isotope signals in firn and ice cores.

Table 2.5: Compilation of stable isotope models on different scales, their advantages and limitations.

Model type	Advantages	Limitations
Lagrangian Rayleigh models	Simplicity, intuitiveness	Initialisation, no vertical processes, parameterisations
2D column models	Vertical process understanding, high temporal resolution	Initialisation, data for validation
2D advection models	Simplicity, reanalysis input data	No process understanding, only one vertical layer
3D GCMs	Global coverage, incorporate available knowledge	No process separation, resolution, parameterisations, model climate
Intermediate approaches	Reanalysis input data, high spatio-temporal resolution, combines GCMs and Lagrangian Rayleigh models	Calculation time, no convection and rainwater evaporation/equilibration

Der Wind bläst, wo er will,
und du hörst sein Sausen wohl;
aber du weißt nicht, woher er kommt und wohin er fährt:
So ist es bei jedem, der aus dem Geiste geboren ist!
Joh. 3, 8

Part I

Lagrangian approach

Chapter 3

A Lagrangian Moisture Source Diagnostic

Motivated by the promises of a Lagrangian moisture diagnostic, as laid out in Chapter 1 and 2, this chapter presents a methodology for identifying moisture transport to Greenland based on backward trajectories. In a Lagrangian framework, the movement of air parcels through space and time can be described by trajectories. The changes in moisture content of such an air parcel along its trajectory will predominantly reflect the effects of precipitation and evaporation processes. This is the basic concept which will be adopted for the method developed here.

Hereby, we pursue two aims: (i) to identify where moisture enters an air parcel, and (ii) to estimate how large the contribution of a source area is to the precipitation from the air parcel at a specific target area. In order to link the moisture source diagnostic to stable isotopes, as presented in the previous chapter, the plateau region of the Greenland ice sheet is chosen as a first target area. The North Atlantic Oscillation (NAO), an important mode of northern hemispheric climate variability, is thereby employed to examine the influence of large-scale circulation changes on moisture sources and transport to Greenland. A large number of stable isotope records from ice cores exists in that region, which allows for a verification of the modelling results.

First, the detailed method for diagnosing moisture sources along a trajectory will be laid out, followed by the calculation setup, some method validation, and a discussion of the identified moisture sources for Greenland during winter. In the chapter thereafter, the method will then be exploited further, and in combination with a stable isotope model interpreted with respect to the inter-annual variability of stable isotope signals in Greenland during winter.

3.1 Identification of moisture uptake

Moisture changes in an air parcel during a certain time interval ($\Delta q/\Delta t$) are generally the net result of evaporation (E) and precipitation[1] (P) (James et al. 2004; Stohl and James 2004):

$$\frac{\Delta q}{\Delta t} = E - P \ \ (g\,kg^{-1}\,6\,h^{-1}). \tag{3.1}$$

As all changes are evaluated on a 6 h time interval, in the following the denominator Δt is dropped for simplicity. Under the assumption that during a 6 h time interval one of the processes dominates over the other, locations of either evaporation or precipitation can be identified along a trajectory (James et al. 2004). Fig. 3.1 shows a sketch of the backward trajectory of an air parcel that is transported from the Atlantic ocean to Greenland. In correspondence with the order of the backward calculation, the arrival point of the trajectory is in the following referred to as the *start* point ($t = 0\,h$), while the earliest point is called *end* point ($t = -54\,h$). The Δq bars above the blue line denote moisture increase ($\Delta q > 0$), while a bar below the line indicates moisture decrease ($\Delta q < 0$) that occurred during a time interval. This implies that Δq is calculated as in Eq. 3.2:

$$\Delta q = q^{t} - q^{t-6\mathrm{h}}. \tag{3.2}$$

To identify an evaporative moisture source that provides the moisture which enters an air parcel during a moisture uptake event along a trajectory, it is in addition required that the moisture increase takes place within the boundary layer (BL). One therefore has to adopt a temporarily relaxed Lagrangian concept, assuming that when an air parcel enters the boundary layer, its surface become permeable, and turbulent fluxes can mix moisture into the air parcel. Hence, if moisture increase of an air parcel is identified inside the boundary layer, a moisture source is identified at this location (Fig. 3.1, ①).

Conversely, if moisture increase is diagnosed outside of the BL, it is not possible to assign this moisture to a specific evaporation source region (Fig. 3.1, ③). It must then be assumed that other reasons caused the moisture increase, such as convection, precipitation evaporation, sub-grid scale turbulent fluxes, numerical diffusion, numerical errors associated with the trajectory calculation, or physical inconsistencies between two ECMWF[2] analysis time steps.

As the interest here is to identify the origin of precipitation in a target area, only air parcels which are precipitating at the time of arrival are traced backward in time (Fig. 3.1, ④). In agreement with the parametrisations of the ECMWF model, it is assumed that clouds exist and precipitation falls whenever a relative humidity threshold

[1] Actually, we diagnose also the effects due to convection (C), diffusion (D), and numerical errors (F), hence $\Delta q = E - P - C - D - F$.

[2] European Centre for Medium-Range Weather Forecast

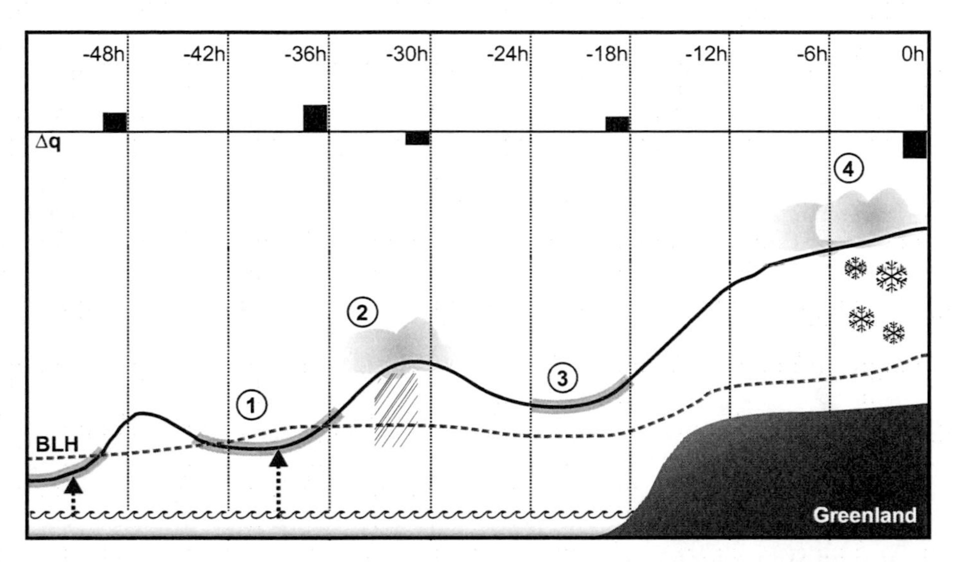

Figure 3.1: Sketch of the method for identifying uptakes along a backward trajectory of an air parcel on the way from the Atlantic ocean to Greenland (blue line). Top row: time before arrival. Δq: changes in moisture content of an air parcel. BLH: boundary layer height. Thickened sections along the trajectory denote sections of moisture increase. See text for details.

of 80% is exceeded (see Sec. A.4.2). The amount of precipitation at the target area is diagnosed from the decrease in Δq during the last 6 h time interval, and then projected onto the arrival location[3]. This follows the ideas of Wernli (1997), and has been applied in several other Lagrangian studies (e.g. James et al. 2004; Sodemann et al. 2006).

In order to derive an estimate of the precipitation at sufficiently high spatial resolution, the air mass over the target area (target volume) is discretised vertically and horizontally into a large number of air parcels (see Sec. 3.3). Under the assumption that all moisture decrease in an air parcel during a 6 h time interval is due to precipitation, and that this precipitation falls immediately and without microphysical processes or interaction with lower levels, the precipitation at the surface P_{surface} is given by the sum of moisture decrease over a whole column of air parcels:

$$P_{\text{surface}} = \frac{1}{g} \sum_{k=1}^{k_{\text{top}}} \Delta q_k \cdot 10^{-3} \cdot \Delta p \;\; (\text{mm}\, 6\text{h}^{-1}), \tag{3.3}$$

where g is the acceleration due to gravity, k is the vertical index, and Δp is the vertical extent of an air parcel.

In detail, the steps taken for the identification of moisture sources are:

[3]it would be more realistic to distribute the precipitation along the path of the air parcel during those 6 h, see Sec. 3.6.1

1. Consider all trajectories that are precipitating (RH>80%) at the time of arrival.
2. Calculate a precipitation estimate for the arrival location according to Eq. 3.3.
3. Trace the air parcel backwards until a positive Δq larger than a threshold $\Delta q_c = 0.2\ \mathrm{g\,kg^{-1}}$ is detected. This threshold helps both to suppress spurious uptakes due to numerical noise, and to keep the analysis computationally feasible (see Sec. 3.6.1).
4. Check if the moisture increase is within the BL. This is the case if the boundary layer height (BLH, in m) predicted by the ECMWF model is larger than the estimated height of the air parcel, assuming a US standard atmosphere:

$$1.2 \cdot \mathrm{BLH} \geq 8000 \cdot \ln(1014/p)\ \mathrm{(m)}, \tag{3.4}$$

   where p is the air parcel's pressure value. In the ECMWF model, the BLH is calculated as a prognostic field from a combined Richardson number and parcel rise method (Troen and Mahrt 1986). To take into account spatial and temporal variability, the BLH and the air parcel altitude are averaged in time at the mean air parcel location during t and $t - 6\,h$ before applying Eq. 3.4. The factor 1.2 is adopted to take into account that moist detrainment may occur at the top of the boundary layer, and that the BLH parametrisation has some uncertainty assigned to it.
5. If the moisture uptake takes place below BLH, a moisture uptake location is identified. The Δq for this location (averaged again in time at the intermediate position) is stored, and several other parameters may be extracted at this location which are important for the further analysis (see Chapter 4). If the moisture increase occurs above the BL, no specific moisture uptake location can be identified, but the location and amount of the above-BL moisture increase are stored for method evaluation purposes.
6. The identification of moisture uptake locations is continued backward in time, either until the trajectory falls almost dry due to rainout ($q \leq 0.05\ \mathrm{g\,kg^{-1}}$), or the end point of the trajectory is reached. One backward trajectory can therefore be assigned with several moisture uptake locations.

With this algorithm, aim (i) can be achieved, namely to identify the locations of moisture uptake of an air parcel along its trajectory.

3.2 Moisture source attribution

In order to reach aim (ii), namely to quantitatively estimate the contribution of a specific moisture source to the precipitation at arrival of the air parcel, the whole transport history of an air parcel has to be analysed chronologically. Over the course of several days, an air parcel may undergo multiple cycles of evaporation and precipitation. Due to rainout, earlier uptakes of moisture will be less and less important for the composition of the precipitation at the arrival site, while later ones weight stronger. Hence, the precipitation at the target area is not simply the sum of the previous uptakes, but only partly attributable to the previous uptake locations. A method is described here which allows to calculate the extent to which each uptake location along a trajectory contributes to the precipitation at arrival.

After all uptake locations are identified according to the method described in Sec. 3.1, their contributions to the precipitation at the arrival site are calculated incrementally, starting at the end point of the trajectory. An example of this procedure, which corresponds to the trajectory displayed in Fig. 3.1, is provided in Table 3.1. Two moisture uptakes occur within the BL (-36 h and -48 h), and one above the BL (-18 h) for this trajectory. The moisture source attribution algorithm proceeds now along the following steps:

1. Initialise all uptake locations with their contribution to the moisture in the air parcel at the time of uptake, Δq^0 (in the example, these amount to $0.8\,\mathrm{g\,kg^{-1}}$ at -48 h and $1.5\,\mathrm{g\,kg^{-1}}$ at -36 h, see Table 3.1).

2. Initialise all $\Delta q = \Delta q^0$

3. From the end to the arrival of the air parcel, evaluate:

Table 3.1: Attribution of moisture sources to the target area precipitation along the trajectory in Fig. 3.1. See text for details.

Time (h)	q ($\mathrm{g\,kg^{-1}}$)	Δq^0 ($\mathrm{g\,kg^{-1}}$)	Δq	f	f_{abl}	f_{tot}
0	2.1	-0.5	-	-	0.11	0.81
-6	2.6	-	-	-	0.11	0.81
-12	2.6	-	-	-	0.11	0.81
-18	2.6	0.3	0.3	0.11	0.11	0.81
-24	2.3	-	-	-	0.00	0.92
-30	2.3	-0.2	-	-	0.00	0.92
-36	2.5	1.5	1.5, 1.380[2]	0.6, 0.53[3]	0.00	0.92
-42	1.0	-	-	-	0.00	0.80
-48	1.0	0.8	0.8, 0.736[2]	0.8, 0.32[1],0.28[3]	0.00	0.80
-54	0.2	-	-	-	0.00	0.00

[1]Updated after later uptake
[2]Discounted after precipitation
[3]Updated after uptake above the boundary layer

- At an uptake location n, calculate the fractional contribution of the uptake amount Δq_n to the moisture in the air parcel q_n as

$$f_n = \frac{\Delta q_n}{q_n}. \tag{3.5}$$

Recalculate the fractional contributions of all moisture uptakes at previous times m with respect to the new moisture content:

$$f_m = \frac{\Delta q_m}{q_n}, \quad m < n. \tag{3.6}$$

In the uptake example, this reduces the contribution of the uptake at -48 h from 80% to 32% of the moisture content at -36 h. The total accounted fraction $f_{tot} = \sum f_n$ amounts then at -36 h to a total of 92% (Table 3.1).

- At a precipitation location, discount all previous contributions to the moisture in the air parcel in proportion to the precipitation amount Δq_n^0:

$$\Delta q_m = \Delta q_m + \Delta q_n^0 \cdot f_m. \tag{3.7}$$

In the example, at -30 h, $-0.2\,\mathrm{g\,kg^{-1}}$ of moisture precipitate from the air parcel. The known sources' moisture amounts in the air parcel are discounted then according to their contribution to the total moisture ($-0.064\,\mathrm{g\,kg^{-1}}$ for -48 h and $-0.120\,\mathrm{g\,kg^{-1}}$ for -36 h, see Table 3.1).

- At an uptake location above the BL, perform the same steps as for an uptake in the BL, except that the moisture increase stems from an unknown source. In the example, at -18 h, $0.3\,\mathrm{g\,kg^{-1}}$ of moisture enter the air parcel at a location above the BL (comp. Fig. 3.1, ②). The contribution fraction f amounts to 0.11, and the previous uptakes are discounted accordingly to $f = 0.53$ and $f = 0.28$.

4. At the arrival location, the sum of the latest fractional contributions of all uptake points, f_{tot}, gives the fraction of the total precipitation to which sources can be attributed. Identifying this fraction is important for evaluating the representativeness of the moisture source attribution. In the example, this amounts to $f_{tot} = 0.81$; $f_{abl} = 0.11$ originate from increases above the BL (Table 3.1). The remaining 8% of water vapour were present in the air parcel prior to the earliest uptake identified, and cannot be attributed to a specific moisture source.

5. Finally, a fractional contribution threshold $f \geq 5\%$ is applied. All values below that threshold are discarded, which considerably shortens the time for subsequent data analysis, while all relevant attributed sources are retained.

In summary, this moisture source attribution method considers the transport history of an air parcel, and discounts earlier contributions depending on later moisture uptakes and precipitation *en route*. The combined methods of moisture source identification and attribution provide information on (i) to what extent moisture sources can be identified (f_{tot}), i.e. the representativeness of our method, and (ii) the weight of each source region, which represents its actual influence (f) on the arrival location precipitation. This is an important prerequisite to weight also the influence of other parameters when considering the isotopic composition of precipitation (Chapter 4).

3.3 Setup of the calculations

A calculation setup was designed, which is suitable for identifying the variability of moisture transport to Greenland corresponding to different NAO phases. The two main requirements of such a setup are that (i) NAO variability should clearly be present in the selected calculation period, and (ii) the spatial resolution of the calculations should be sufficiently high to identify regional differences in moisture transport to Greenland.

As the NAO has its active phase in Northern Hemispheric winter, and phases typically persist on time scales of several weeks to one month, we selected 30 winter months from the time period 1958–2002 (which is the range of the ERA40 dataset). 10 months each were chosen where the NAO indices of Hurrell (1995) are clearly positive (3.89±0.60, NAO+), negative (-5.00±0.83, NAO−), and neutral (0.01±0.30, NAO=) (Table 3.2). Care was taken that the selected months did not fall into a strong ENSO (El Niño/Southern Oscillation) episode, which could, via teleconnections, introduce additional variability. The chosen subset, though necessarily limited, should hence represent the typical range of NAO variability.

Table 3.2: Case names, corresponding NAO index (Hurrell 1995), and percentage of precipitating trajectories for the 30 chosen months.

NAO positive			NAO neutral			NAO negative		
Index	Case[1]	Precip[2](%)	Index	Case	Precip (%)	Index	Case	Precip (%)
4.8	199702	1.14	0.5	198101	1.29	-4.0	196902	3.32
4.7	199002	0.97	0.4	198312	1.45	-4.3	196901	2.28
4.3	199001	1.95	0.2	199112	2.85	-4.4	195901	1.21
4.1	198902	0.48	0.2	196801	1.35	-4.5	196601	1.66
4.1	198401	1.14	0.1	198802	2.38	-4.6	198701	6.69
3.9	200002	1.07	-0.1	199302	2.75	-5.1	196502	5.89
3.7	197401	2.42	-0.2	196412	1.96	-5.1	199612	4.53
3.2	199301	1.03	-0.3	199712	3.90	-5.2	199512	3.30
3.1	199502	1.19	-0.3	196501	2.15	-5.8	196112	2.46
3.0	199312	1.65	-0.4	199402	2.35	-7.0	196301	2.91

[1] Format YYYYMM

[2] RH≥80% at $t = 0$

In order to achieve a spatially detailed picture of the moisture transport to Greenland, the atmosphere above the ice sheet was discretised horizontally (60×60 km) and vertically ($\Delta p = 30\,\mathrm{hPa}$) from the surface to a pressure level of 480 hPa into air parcels of equal (dry) air mass. Starting points of trajectories were selected over the Greenland plateau, which we defined as all locations above 2000 m altitude (Fig. 3.2).

For each of the 30 months, every 6 h, starting at each of the starting points above the plateau region, 20-day kinematic backward trajectories were calculated if RH was ≥80% at the respective starting point. Trajectories were calculated with the Lagranto model (Wernli and Davies 1997). Lagranto used the 3-dimensional 6 h-ly ERA40 wind fields (u, v, ω) for the calculation of the backward trajectories, and stored latitude, longitude, pressure, potential temperature, and specific humidity of the traced air parcels at a 6 h interval. Infrequently, calculations were halted when an air parcel left the calculation domain (i.e. the Northern Hemisphere). When air parcels intersected with the orography, they were displaced several hPa above the surface, so that calculations could be resumed (Lagranto's j-flag).

With this setup, ~700000 trajectories could potentially be calculated for every month. However, for the 30 months studied here, typically only about 1–6% of these fulfilled the RH criterion (Table 3.2). More than 95% of the precipitating trajectories were associated with at least one moisture uptake location (Appendix B). Some calculation and data file errors were encountered, but these were far too few to affect the results noticeably.

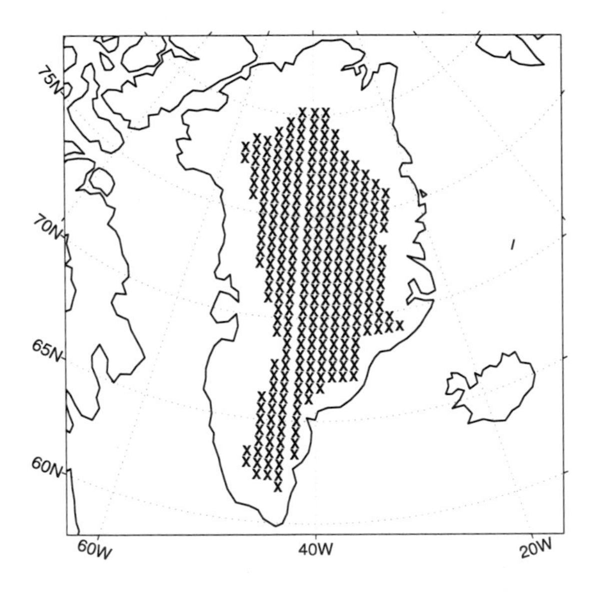

Figure 3.2: Horizontal discretisation of the air mass above the Greenland plateau into 284 columns of starting points for the calculation of backward trajectories.

3.4 Method validation

Before presenting the results of the moisture source analysis, an attempt is undertaken to assess the soundness of the calculation results. Issues considered include the orographic representation of Greenland, the amount and variability of the diagnosed precipitation, and several transport-related parameters.

An important aspect of modelling atmospheric transport to Greenland is the extent to which the ERA40 orography resembles the actual orography of the island. Particularly important features for the flow are the maximum altitude and the separation into a flat plateau region and a steep slope (Schwierz 2001). Comparing the orography of Greenland in the (spectral) ERA40 model with a NOAA dataset of the same resolution (Fig. 3.3), an overall high degree of consistency can be noted. The maximum altitude of the ERA40 orography is above 3000 m, and is separated into a flat plateau and a steep slope. The slope is less steep in the ERA40 than in the NOAA orography, the area of maximum altitude is shifted somewhat to the northwest, and smaller-scale features are not well resolved. At lower elevations, due to the ERA40 model's spectral numerics, a land mask is required to clearly demark the coast line (Fig. 3.3a, grey shading). Nevertheless, no obvious biases should have been introduced into the calculation results by the orographic representation Greenland.

Next, the precipitation over Greenland derived from the Lagrangian methodology (Eq. 3.3) is compared with the prognostic precipitation from the ERA40 dataset. As $t + 6\,\mathrm{h}$ precipitation forecasts are typically biased by model spin-up, precipitation has been alternatingly calculated from the $t + 12 - 6\,\mathrm{h}$ and $t + 18 - 12\,\mathrm{h}$ forecasts from pre-

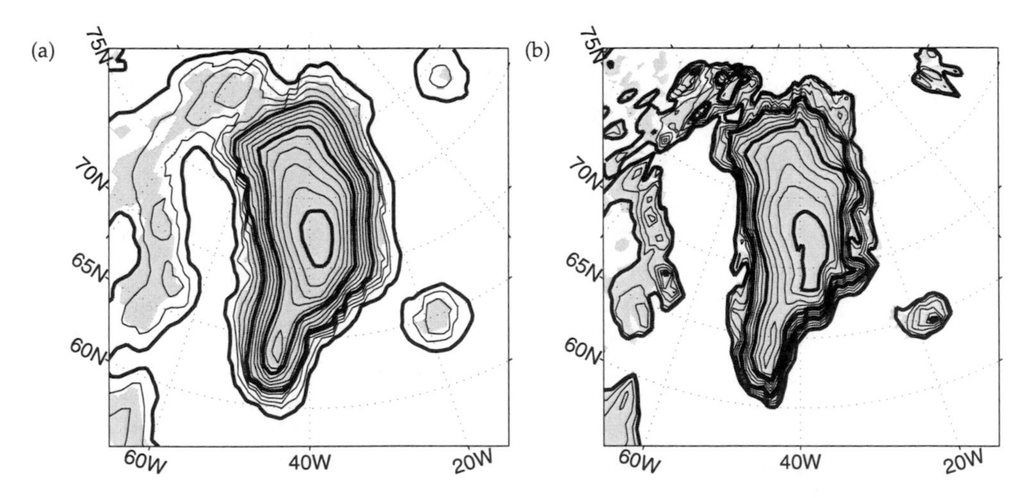

Figure 3.3: Orography of Greenland from (a) the ERA40 model (spectral resolution T159 interpolated on a 1°×1°grid), (b) a NGDC-NOAA topography data (1°×1°grid). Contour interval is 200 m, for thick contours 1000 m. First contour is at 50 m.

ceding simulations. A high degree of correspondence between the two precipitation estimates is apparent (Fig. 3.4a, b). The NAO variability is very similar for both estimates, with precipitation limited to the eastern flank during NAO+ months, and a shift to the northwest and an overall increase of precipitation magnitude during NAO−. The phase mean for NAO= exhibits a pattern similar to NAO−, but of intermediate magnitude (not shown).

In more quantitative terms, the Lagrangian precipitation estimate is biased high by ~20–40%, in particular in the regions of high precipitation at the south-eastern slope (Fig. 3.4c). This high bias can be partly explained by the assumption that the total Δq during the last 6 h before arrival falls as precipitation over the arrival location, instead of partial rainout before (see Sec. 3.6.1). With southerly (easterly) advection, this leads to a northerly (westerly) shift of the precipitation estimate. Furthermore, other studies using the same or more detailed Lagrangian precipitation estimates also found a general high bias (Wernli 1997; Eckhardt et al. 2004; James et al. 2004). Despite these limitations, the simplification that precipitating air moves slow enough so that precipitation can be projected forward onto the arrival location is considered here as a valid first order approximation, in particular over the drier interior of Greenland.

While the correspondence between the ERA40 precipitation and the Lagrangian estimate can generally be considered as reasonable, the question remains how the modelled precipitation compares with observations. Reliable precipitation measurements are difficult to obtain for Greenland. Bromwich et al. (1999) diagnosed the precipitation over Greenland from ECMWF analysis data, and found precipitation patterns for NAO+ and NAO− means that closely resemble Fig. 3.4a. Validating the variation of snow accumulation in Greenland from ECMWF analyses against ice-core data, Hanna et al. (2001) found, despite a 20–30% low bias, reasonable first-order agreement. Studying the NAO variability of Greenland precipitation, Appenzeller et al. (1998a) found good correspondence between ERA15 precipitation and ice accumulation data, with higher accumulation along the east coast (west coast) during the NAO+ (NAO−) phase. Similar studies for the ERA40 dataset are at present not available (A. Simmonds, *pers. comm.*, 2005), but due to model improvements it is expected that results should further improve.

The precipitation estimate can also be used to assess the representativeness of the selected cases. The increasing tendency of the monthly mean precipitation over Greenland with decreasing NAO index is also evident in Fig. 3.5. NAO positive cases cluster near low monthly means (group mean 6 mm), NAO negative cases have a considerably larger variability and a 3 times larger group mean (19 mm). NAO neutral months take an intermediate position, both with respect to variability and group mean (14 mm). The selection of cases hence shows a consistent picture, and no outliers in terms of precipitation have been included. January 1987 has the largest monthly mean, but will not dominate the analysis of the NAO negative group.

Further insight into the performance of the method can be gleaned from the inspection of several parameters related to the identified arrival and uptake locations of

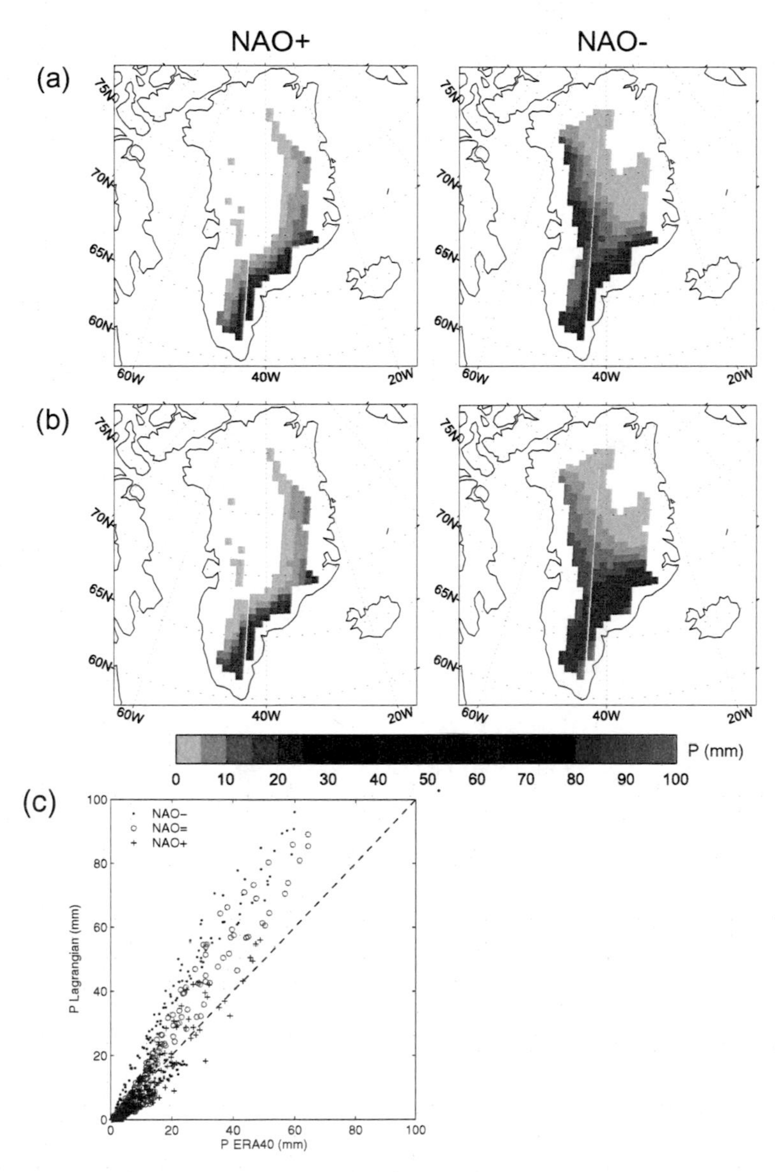

Figure 3.4: Phase mean monthly accumulated precipitation over Greenland (mm). (a) ERA40 precipitation gridded to the calculation starting point grid, (b) Lagrangian precipitation estimate, (c) correlation between ERA40 and Lagrangian for the three NAO phases with a 1:1 reference line (dashed).

moisture. Fig. 3.6a shows a histogram of the pressure level at which precipitating air parcels arrive over Greenland, weighted by the respective precipitation amount. The mean arrival pressure of precipitation is at ~660 hPa (~3200 m asl), more than 70% of

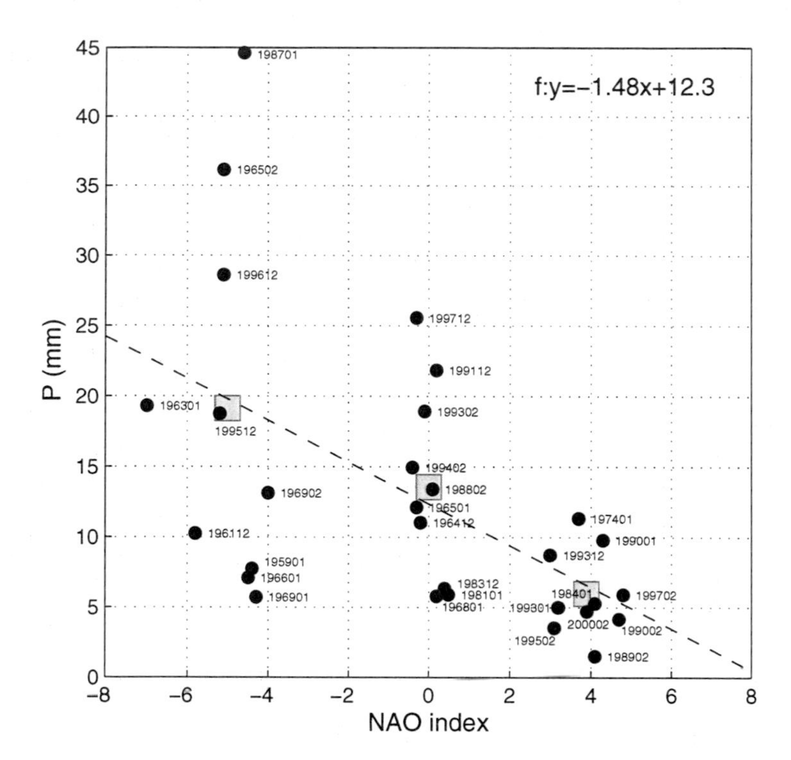

Figure 3.5: Scatter plot of the mean accumulated monthly precipitation over the Greenland plateau vs. the NAO index for the 30 selected months. Dashed line is a linear regression, boxes denote group means.

the precipitation arrives between 720–600 hPa, with a tendency towards arrival at lower pressure during NAO+ months (red dashed line) than during NAO negative months (black line). NAO= months take intermediate position (gray shading), but are more alike to the NAO− phase.

Fig. 3.6b shows the fraction of precipitation which can be attributed to specific source regions (f_{tot}, see Sec. 3.2), weighted by the respective precipitation amount. From this methodology, for half of the Greenland precipitation, $\geq$60% (median value) of the arrival precipitation can be attributed to specific moisture sources. In total, $\sim$59% of the moisture origin can be attributed to moisture sources. About 5% of the precipitation events reach more than 90% attribution, while $\sim$5% have $\leq$5% attribution. This result is rather invariant with NAO phase, which indicates that the extent of source attribution is comparable for all NAO phases.

The accounted fraction is directly related to the number of uptake events per trajectory. Fig. 3.6c shows that on average 3–6 events (out of possible 80) occur per trajectory, while some have more than 20 uptake locations diagnosed. A slight tendency towards

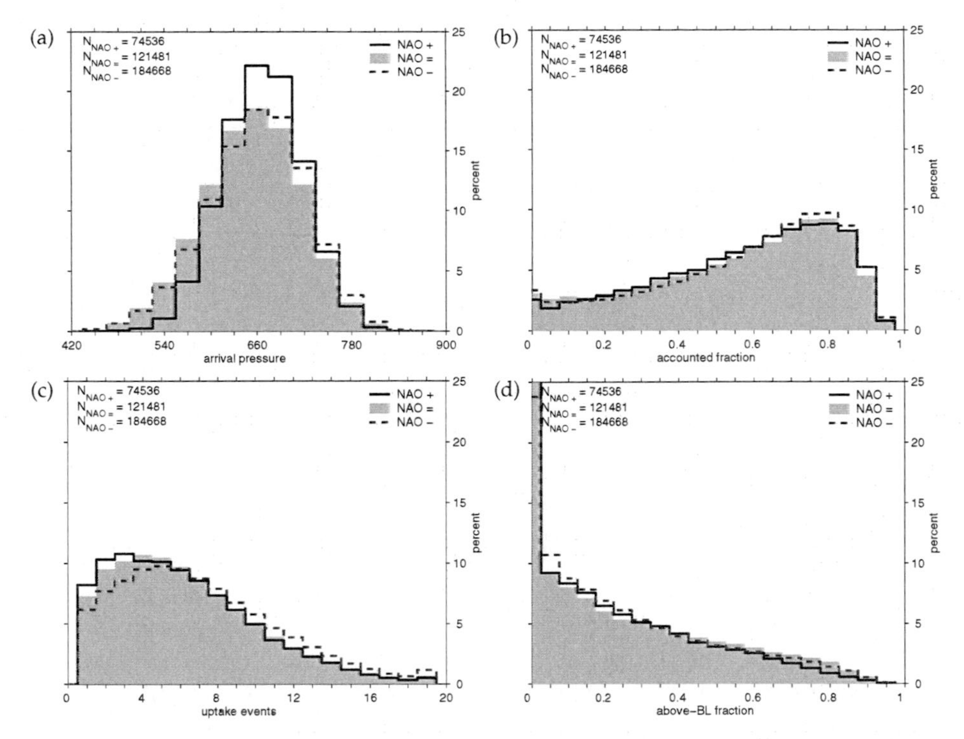

Figure 3.6: Method validation histograms for all trajectories that transport moisture to the Greenland plateau: (a) arrival pressure, (b) accounted fraction, (c) uptake events per trajectory, (d) above-BL uptake fraction. Histograms (a) and (b) are weighted by the precipitation amount at arrival. N designates the number of air parcels which contributed to the respective phase. Gray area: NAO neutral phase; black solid contour: NAO positive phase; red dashed contour: NAO negative phase.

more uptake points for the negative NAO phases can be noted.

In addition, a substantial number of tests was undertaken to evaluate the response of the method to the BL criterion (Eq. 3.4) and the two uptake threshold values.

As noted in Sec. 3.1, the locations where moisture increase was detected above the boundary layer were also stored and saved. The histogram of these above-boundary layer (ABL) uptakes (Fig. 3.6d) shows that about 60–70% of the trajectories are associated with ABL uptakes. Only ~5% have more than 80% of ABL moisture. The overall mean is 26%, with slightly lower fractions towards NAO+ months. As is shown in Fig. B.2, the ABL uptakes mostly overlap with the general moisture source regions (see below), which further confirms the results of the Lagrangian moisture source diagnostic. The remaining ~16% of moisture then originate from completely unknown source regions.

Two arbitrary thresholds are included in the method: (i) the minimal positive Δq required to diagnose a moisture uptake event, and (ii) the minimal percent contribution of a moisture uptake location to the total moisture in an air parcel. The sensitivity of the method to these thresholds has been tested for one month, and the presently used values have been chosen as a tradeoff between identification of the dominant moisture sources, exclusion of spurious moisture uptakes, and performance of the overall calculations and analysis. Choosing a smaller Δq_c only led to a slight increase of f_{tot} beyond the current values. Still, some subjectivity remains in choosing these parameters, and other study areas or seasons may require different threshold values.

3.5 Moisture source regions

Based on the initial method validation presented above, this section presents the identified moisture source regions for winter precipitation in Greenland[4].

3.5.1 Source regions for Greenland

Fig. 3.7 shows the phase mean moisture sources for Greenland precipitation. Only the contribution of the evaporation at the uptake locations to the precipitation over Greenland is shown. The pattern can hence be interpreted as a Lagrangian backward projection of the monthly mean winter precipitation in Greenland onto the respective source areas.

The source regions look strikingly different for the three NAO phases. While during the NAO+ phase uptake locations are confined to areas of the Atlantic north of 40°N, a gradual southward extension beyond 30°N occurs towards NAO− months. In addition, a large dominant patch of moisture sources with a maximum near the British Isles appears during NAO− months, which is not present for the NAO+ phase. For all NAO phases, almost all moisture uptake locations are in the North Atlantic, with only small contributions from the Mediterranean and the Gulf of Mexico during the NAO− phase.

A comparison with the moisture uptake locations without taking into account the source attribution shows a more southerly location of the uptake maxima (Appendix B). This highlights how rainout during the transport decreases the significance of more distant moisture sources. Additional insight is gained from a consideration of the moisture sources for different arrival sectors of the Greenland ice sheet (Sec. B.4). Different sectors of the ice sheet obviously experience differing changes in the main moisture source with the NAO.

A more quantitative view of the changes in moisture source areas is provided by dividing the Northern Hemisphere source areas into several sectors (Fig. 3.8a). The ocean

[4]It is important to note in advance that the diagnosed moisture sources are representative only for water vapour that will precipitate over Greenland. The information given here is hence different from the overall picture of evaporation in the North Atlantic.

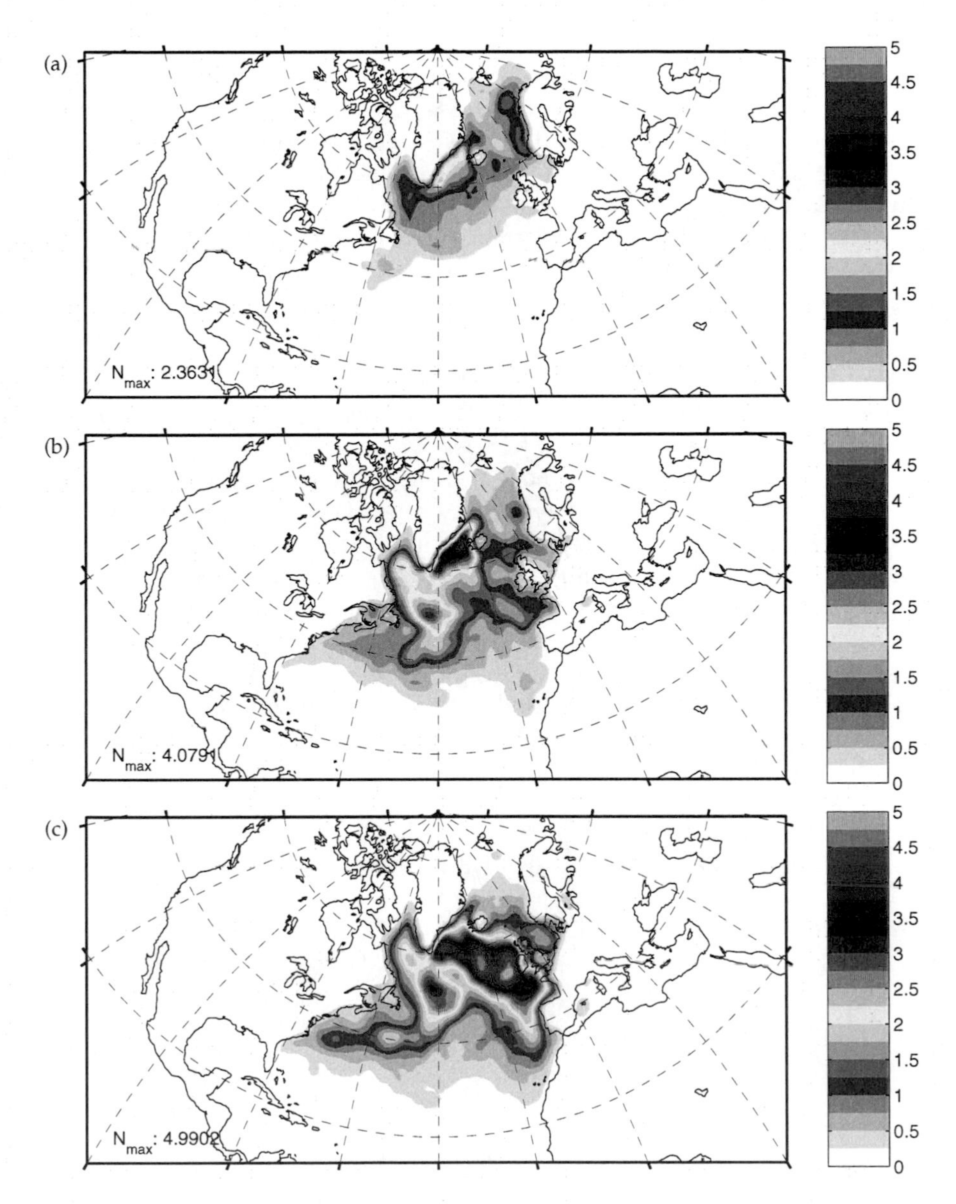

Figure 3.7: Phase mean moisture sources for (a) NAO+, (b) NAO=, (c) NAO− months. Uptake locations show the contribution to precipitation in Greenland. Units are mm precipitation contribution integrated over 10^4 km^2.

sources are divided into four North Atlantic sectors (northeast, northwest, southeast, southwest), an Arctic sector, the Mediterranean, and the Pacific. Furthermore, it is distinguished between moisture uptakes over land and over sea. The percent contribution of the sources changes strongly for the three NAO phases (Fig. 3.8b). The southeastward shift from NAO+ to NAO− manifests itself in the decrease of Arctic ocean uptakes from 50.9% to 19.9%, along with an increase of the northeast Atlantic contribution (19.1%→38.4%), and the southeast and southwest Atlantic contributions (2.9%→9.2% and 0.8%→8.4%). Contributions from the Mediterranean and other sources, in particular the Pacific, can hardly be identified. Also, all land sources jointly contribute only about ~0.2% to the winter-time precipitation in Greenland, which underlines the solitary role played by the North Atlantic ocean for the moisture supply of the Greenland ice sheet during winter. A regional picture of the moisture sources for different areas of the Greenland ice sheed is provided in Appendix B.

In the conceptual model of Johnsen et al. (1989), a fixed subtropical moisture source south of 40°N was assumed in the western North Atlantic. It is evident from the results presented here that this *ad hoc* model does not correspond to diagnosed moisture sources, even though arguments for the plausibility of that model have been put forward (White et al. 1997). The moisture sources derived from the present method can furthermore be compared to the results of tagging GCM studies. In the winter period of a present climate simulation using the GISS GCM, Charles et al. (1994) found 23% contribution from the Arctic Ocean and the Norwegian-Greenland Sea, which corresponds well with the Arctic sector of this study. Their North Atlantic sector (30°–50°N) contributed 31%, which is only about half the total of the NW and NE sectors here. This may be partly attributable to the more southerly location of their moisture source. Their Tropical Atlantic source (30°N–30°S) contributed 11%, again corresponding reasonably to our combined SW and SE sectors. The major difference is a 16% contribution of the Pacific to precipitation over Greenland in the work of Charles et al. (1994), compared to 0% for this study. A combined isotope/tagging study by Werner et al. (2001) with the ECHAM model (resolution T30≈3.75°x3.75°) confirmed these earlier findings. Average

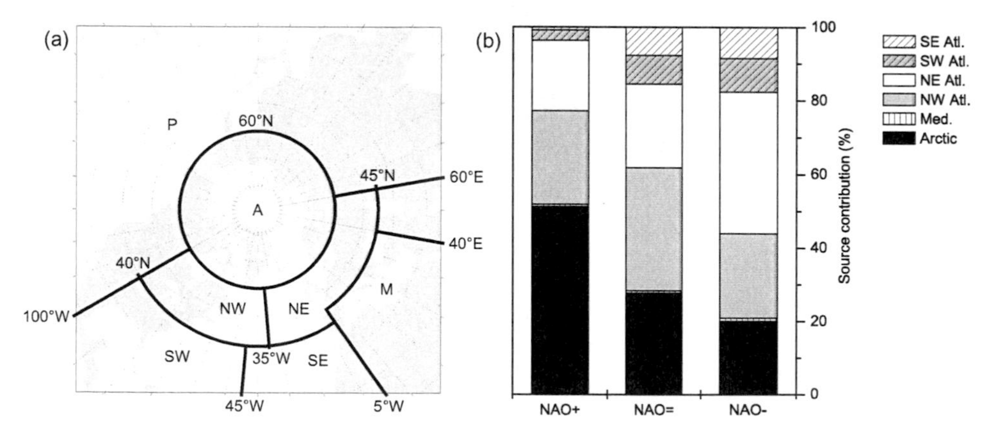

Figure 3.8: (a) Definition of the uptake sectors. A: Arctic Ocean, NW: north-western North Atlantic, NE: north-eastern North Atlantic, SW: south-western North Atlantic, SE: south-eastern North Atlantic, M: Mediterranean, P: Pacific. (b) Relative contributions of the uptake sectors to the transport of moisture to Greenland. No contributions from the Pacific sector are identified.

winter time precipitation for Greenland consisted of 25% Arctic, 40% North Atlantic, 15% tropical Atlantic, 18% Pacific, and 6% continental moisture (Werner et al. 2001). An important limitation of these tagging GCM studies is the grid resolution: the precipitation maximum along the west flank of Greenland is not reproduced by this simulation. The differences for the Pacific source could also be explained with the low resolution of both GCMs, in particular near the poles, which could lead to unrealistically large cross-polar moisture transport to Greenland. Nevertheless, the reasonable correspondence between the Lagrangian findings and those from the completely different GCM approach are very encouraging.

Latitudinal shifts of the uptake regions also vary spatially over the Greenland ice sheet (Fig. 3.9b). Parts of the ice sheet closer to the coast are fed by more northerly

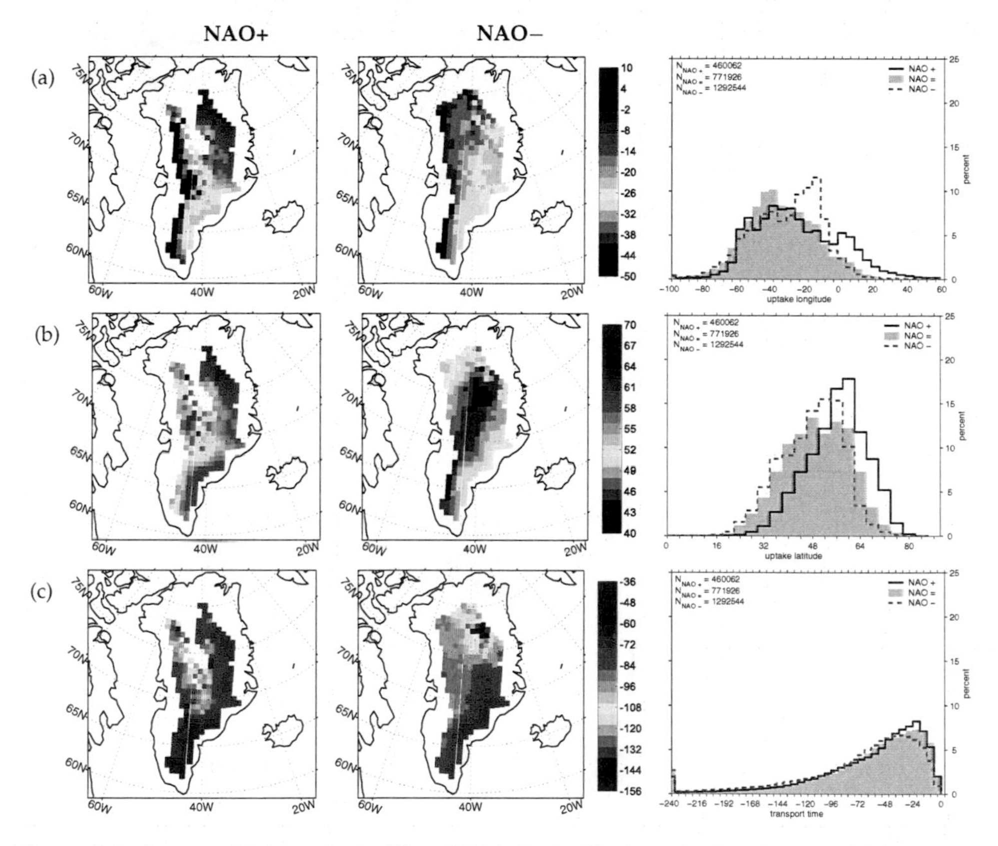

Figure 3.9: Averaged (a) longitude (°) and (b) latitude (°) of uptake locations, and (c) transport time (h) of moisture that contributes to precipitation at a given location on the Greenland plateau. Left and middle panels: Lagrangian forward projections of uptake conditions projected onto the arrival grid over Greenland, weighted by the contribution of each uptake location to the phase mean precipitation over Greenland. Right panels: Histogram of the respective uptake condition.

source regions, while the (higher) interior of the ice sheet receives its moisture from on average 8° latitude more southerly locations. This highlights the orographic influence on the moisture supply to different regions of the ice sheet. For positive NAO months, the source regions are on average shifted ~8° latitude towards more northerly uptakes, compared to the negative and neutral NAO phases (Fig. 3.9b, right panel).

The mean transport time of moisture is calculated as the time between an uptake event and the arrival of the corresponding air parcel over Greenland. Its spatial pattern gives an indication of the time range of moisture transport (Fig. 3.9c). During the NAO+ phase, the average transport time scale is 3–4 days, with areas close to the east coast of Greenland having more immediate moisture transport. This corresponds well with the high latitudes of the uptake locations during this NAO phase. For the NAO− phase, a N–S gradient of transport times is apparent. While the south-eastern coast is also associated with a transport time of 3–4 days, transport to central Greenland takes on average 4–5 days. A region in the very north of Greenland has significantly longer transport times associated (~7–9 days). This could indicate a tendency towards more meridional instead of zonal moisture transport during NAO− months.

3.5.2 Influence of NAO variability

It is obvious from Figs. 3.7 and 3.9 that the NAO and the corresponding circulation changes in the Northern Hemisphere profoundly alter the moisture transport regime to Greenland (Chapter 1 and 2). The implications of these circulation changes for several parameters related to moisture transport are now discussed further (Fig. 3.10).

The mean sea level pressure (SLP) pattern for the months selected here shows a strong pressure gradient for the NAO+ phase, and a weak one for the NAO− phase (Fig. 3.10a, b). This mean picture is very similar to that published by others for the ERA40 dataset (e.g. Scherrer et al. 2006), which confirms again that the chosen cases capture the NAO variability. During the NAO+ phase, moisture sources are confined to the north of the strong pressure gradient (Fig. 3.7a), suggesting that cyclones play a major role for precipitation over Greenland (Bromwich et al. 1999). During the NAO− phase, source regions are clearly less confined (Fig. 3.7c), which corresponds to the much weaker pressure gradient across the North Atlantic (Fig. 3.10b). More meridional transport can take place, which corresponds to the east Atlantic maximum of moisture sources and the further southward extent during NAO− months.

Precipitation averages for the NAO+ phase reflect again the more confined cyclone track (Fig. 3.10g), with maxima near the coasts of Greenland, Norway, and the British Isles. The interior of Greenland remains virtually dry during this phase, as does most of the Canadian Arctic. During the NAO− phase in contrast, precipitation spreads more zonally along a broad swath across the North Atlantic, and also continues into the Mediterranean (Fig. 3.10h). A maximum of precipitation is visible over Portugal. Over Greenland, precipitation extends further into the interior of the ice shield, in particular

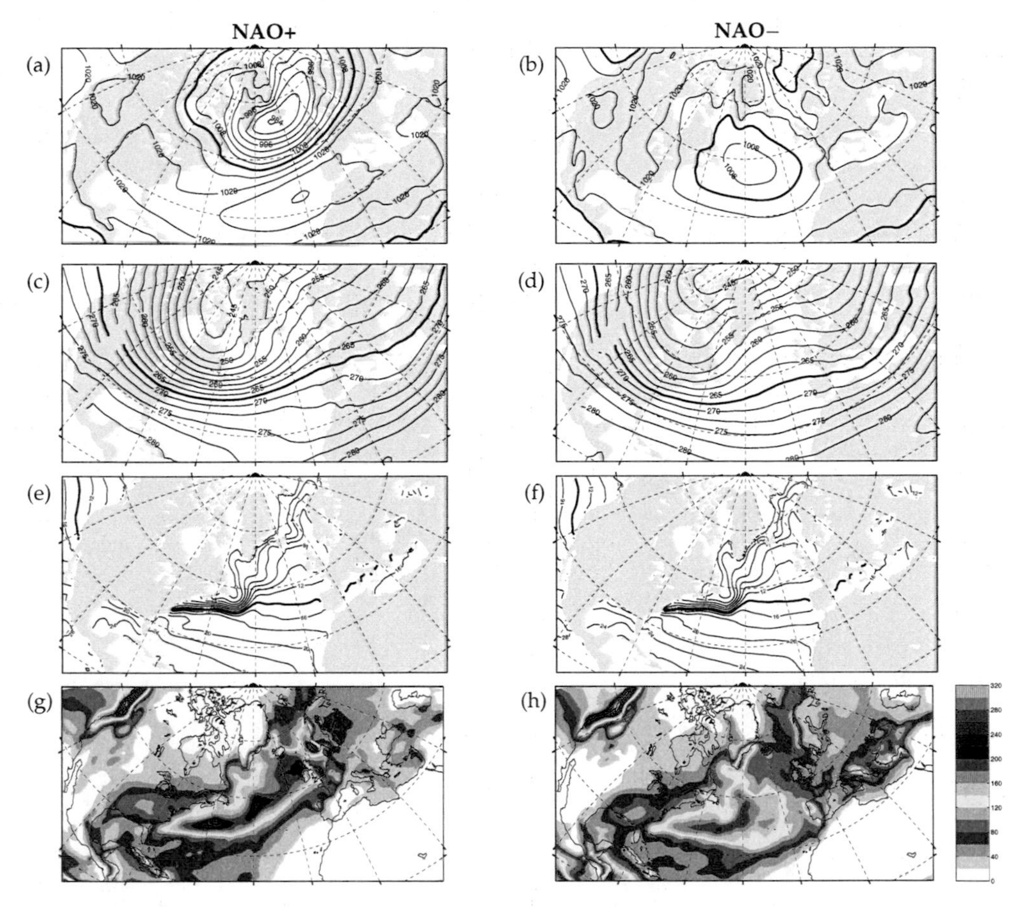

Figure 3.10: Mean conditions during the selected NAO positive and NAO negative months of (a),(b): sea level pressure (contour interval 4 hPa); (c),(d): temperature at 700 hPa (contour interval 2.5 K); (e),(f): sea surface temperature (°C, contour interval 2 K); (g),(h): monthly accumulated precipitation (contour interval 20 mm).

along the western flank (c.f. Fig. 3.4).

The mean temperature at 700 hPa (Fig. 3.10c, d) shows 2-3 K warmer conditions near Greenland during the NAO− phase, while colder conditions prevail between 30°–50°N over the Atlantic, Europe, and Scandinavia. This opposing temperature signal in Norway and Greenland is sometimes termed the Greenland temperature seesaw (van Loon and Rogers 1978; Barlow et al. 1993). The negative temperature anomaly over Europe during the NAO− phase is closely coupled with more frequent blocking episodes (Scherrer et al. 2006). The atmospheric temperature change is particularly relevant for the isotopic fractionation regime, and will be discussed further in Chapter 4.

Changes in sea surface temperature (SST) with the NAO are relatively subtle (Fig. 3.10e, f). During the NAO− phase, the Labrador sea and the subtropical North

Atlantic show slight warming, while the Arctic sea is somewhat cooler. It is currently still a matter of controversy whether SST changes with the NAO are response to or driver of that climate mode (Hurrell et al. 2003). The evaporation temperature of the moisture sources for Greenland precipitation is certainly more likely to change due to the source region changes than due to local SST changes. In turn, the source region SST changes can have a strong influence on the isotopic composition of the evaporating water vapour, a topic which will be further explored in the next chapter.

3.5.3 Transport and uptake processes

To illustrate the flow patterns associated with the identified source regions, two typical examples for moisture transport to Greenland are examined. The moisture sources for the first example (February 1997) closely resemble the mean NAO+ pattern (Fig. 3.11a). Two maxima of moisture uptake are apparent, one along the Norwegian coast, and the other off south-eastern Greenland. Two coherent trajectory clusters[5] arriving at 00 UTC 19 Feb 1997 are shown in Fig. 3.11b, coloured by specific humidity. A cyclonically curved cluster of trajectories represents air parcels that originated over eastern Canada, descended to the sea surface (Fig. 3.11c, thin lines), increased in specific humidity about 96 h before arrival south of Greenland, and precipitated then while rapidly ascending from 850–900 hPa onto the Greenland plateau (600–700 hPa). A more northerly trajectory cluster represents air parcels that moved from the Arctic at diverse altitudes in south-easterly directions (Fig. 3.11c, thin lines), until 48 h before arrival they sharply changed direction, descended westward to the coast of Norway, increased their moisture content over the Norwegian Sea, and delivered it as precipitation to eastern Greenland the following day. Both trajectory clusters were clearly part of a mid-latitude cyclone that was present east of Greenland at that time (not shown). The cluster of Canadian origin was however incorporated in the cyclone 2–3 days earlier than the cluster from Norway.

The second example (February 1965) has a moisture source pattern that is quite similar to the NAO– mean picture (Fig. 3.12a). Of the months examined here, it is the case with the second-largest monthly mean precipitation (c.f. Fig. 3.5). A particularly strong moisture contribution from the east Atlantic is apparent in that month. One coherent trajectory cluster arriving at 12 UTC 15 Feb 1965 was extracted (Fig. 3.12b, c), using the same clustering methodology as for the previous case. The trajectory cluster originated at mid altitudes (400–600 hPa) over the Arctic sea, and slowly descended anticyclonically towards Spain. Five days before arrival over Greenland, the moisture content of the air parcels increased by $4\,\mathrm{g\,kg^{-1}}$, corresponding to the south-eastern maximum in moisture uptake locations in Fig. 3.12a. The air parcels then moved in north-westerly direction at low altitudes, resembling the moisture flux in *tropospheric rivers* (Newell et al. 1992), and the moist ascending coherent ensembles of trajectories (Wernli 1997,

[5]Trajectory clusters were extracted from a Ward's clustering using the 6 h-ly trajectory co-ordinates throughout the last 5 days before arrival (Sodemann 2000). 8 trajectory clusters were retained.

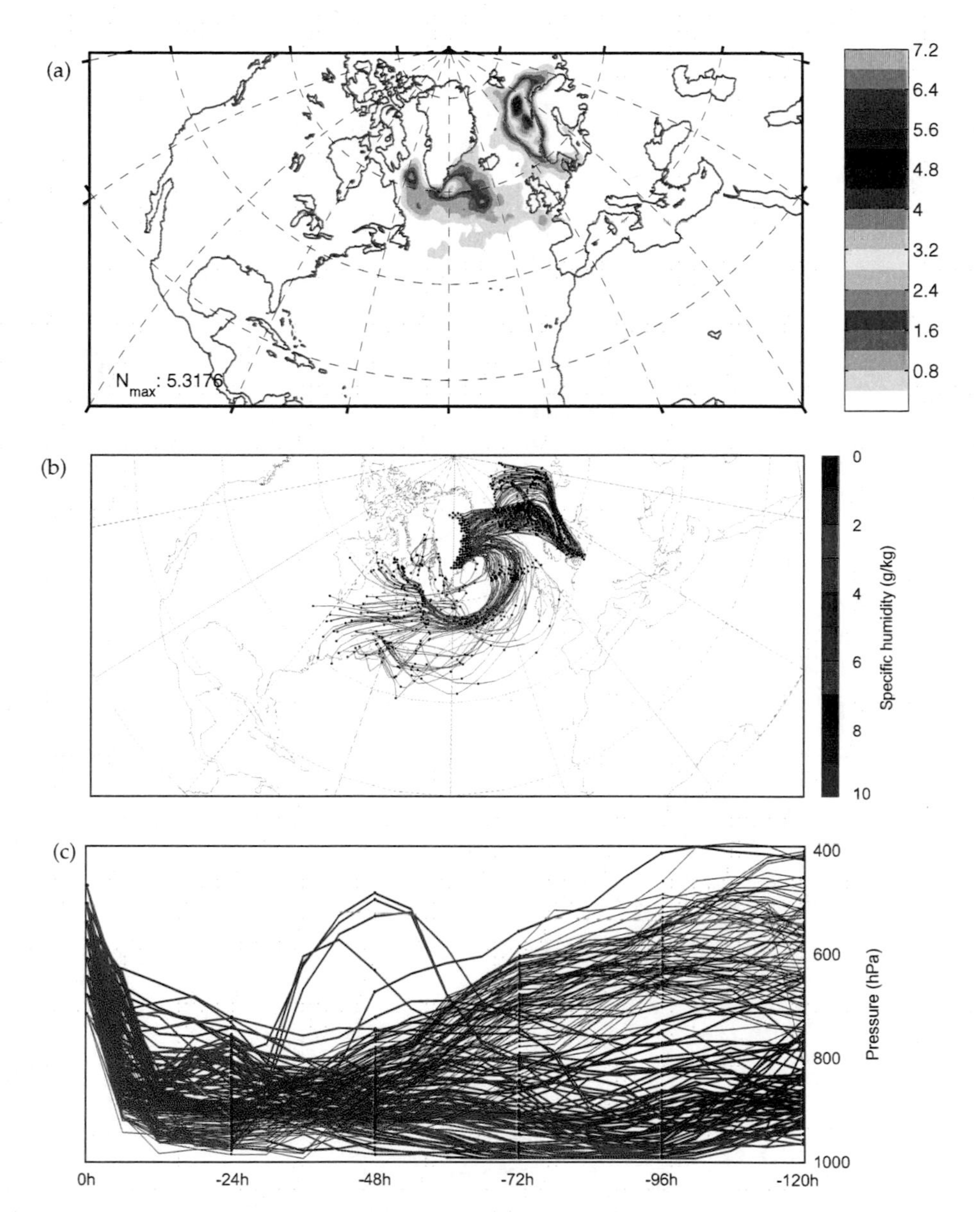

Figure 3.11: Case study for February 1997. (a) Moisture source regions (mm contribution to Greenland precipitation), (b) 7-day back-trajectory clusters, arriving at 00 UTC 19 Feb 1997, (c) Time-pressure diagram of the two trajectory clusters. Thin lines: northerly cluster, thick lines: southerly cluster. Shading in (b) and (c) gives specific humidity (g kg^{-1}), black dots denote 24 h intervals.

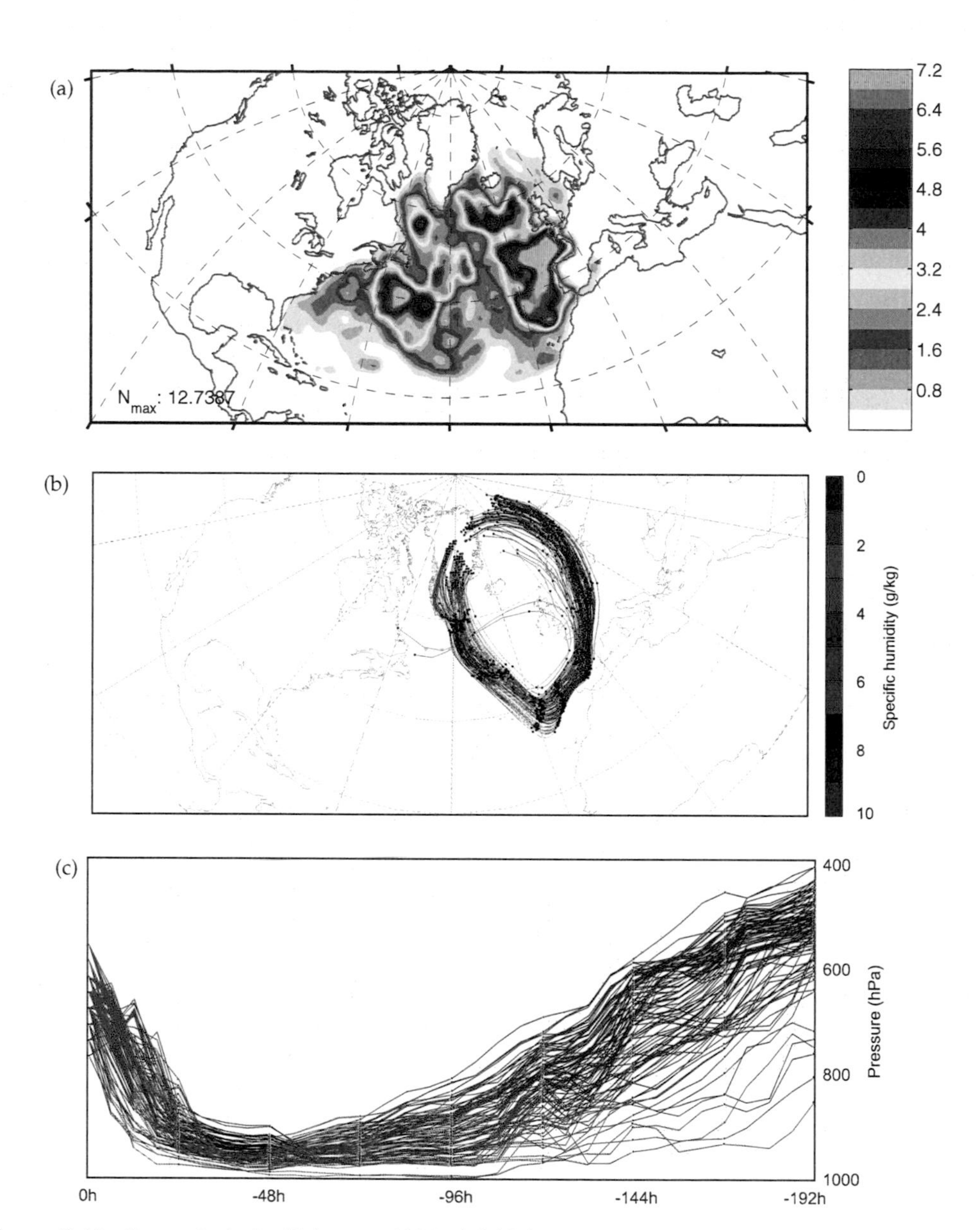

Figure 3.12: Case study for February 1965. (a) Moisture source regions (mm contribution to Greenland precipitation), (b) 8-day back-trajectory cluster, arriving at 12 UTC 15 Feb 1965, (c) Time-pressure diagram of the trajectory cluster. Shading in (b) and (c) gives specific humidity (g kg^{-1}), black dots denote 24 h intervals.

his Fig. 6). Finally, the air parcels ascended from southerly directions onto the Greenland plateau.

The transport patterns examined in these two case studies give more insight into the regional differences in uptake latitude, longitude, and moisture transport time over the Greenland plateau (Fig 3.9). During the NAO+ phase, south-westerly arrivals along the Greenland east coast could correspond to transport similar to the Canadian cluster, while north-easterly arrivals would correspond to transport more alike to the Norwegian cluster (Fig. 3.11). For the NAO− phase, the increasing transport time towards northern parts of the Greenland plateau is in line with more frequent meridional moisture as suggested by the trajectory cluster in Feb 1965 (Fig. 3.12).

Furthermore, these two illustrative examples suggest that cyclones and anticyclones are both very important for the transport of moisture to Greenland, be it by delivering precipitation to the Greenland plateau, or by supplying dry air from subsidence to lower levels. However, it is clear that no representative picture can be derived from these two cases alone. A more climatological view of the role of cyclones and anticyclones by examining the relative position of moisture uptake locations with respect to their centre has recently been initiated, but a clearer picture hat yet to emerge.

3.5.4 Pre-uptake locations

The pre-uptake history of air masses, defined here as the history of an air parcel before its first moisture uptake, is a further issue related to the processes surrounding moisture uptake. Evaporation into an air parcel is driven by the saturation water vapour pressure deficit. Hence, moisture uptake is larger the drier the air, and the larger the temperature contrast between air and water. To investigate the pre-uptake history of air parcels, their first (earliest) uptake location, and the locations 1, 3, and 5 days prior to the first uptake were identified (Fig. 3.13). During the NAO+ phase, the areas of first uptake are distributed along the eastern North American coast, and in the northern Norwegian Sea (Fig. 3.13a, blue shading). With decreasing NAO index, areas of first moisture uptake also gradually appear along the European coastline, and northwestern Africa (Fig. 3.13b, c). One day before the first uptake, the air-parcels are to the most part located over the continents (Fig. 3.13, solid contour). This impression further increases for the locations at 3 and 5 days before the first uptake (Fig. 3.13, dashed contours).

It is interesting to note that the continental origin is a common feature of pre-uptake history, both on the American east coast and the European west coast. While for the American continent, the westerly outflow is in agreement with the mean flow, the European easterly outflow is oriented in the opposite direction, which suggests that mid-latitude disturbances play an active role in these areas. Along the east coast of Greenland, a consistent feature in the locations one day before uptake indicates a possible importance of cold and dry katabatic outflow for evaporation in the Arctic ocean.

Finally, it is interesting to note that during the NAO− phase pre-uptake locations are detected over northern Africa, which could indicate that warm and dry Saharan air is important for moisture uptakes in the eastern North Atlantic maximum (c.f. Fig. 3.7c).

The identified pre-uptake locations fit well into the picture of cyclonic moisture transport gleaned from the interpretation of Figs. 3.11 and 3.12. Land-sea contrast and evaporation are obviously important for cyclone and anticyclone life cycles in the North Atlantic (Schwierz 2001). It appears that cyclones could include moisture at two major locations during their North Atlantic life cycles in NAO− winters: first, during cyclogenesis over the west Atlantic when cold continental air is swept from the North American land mass, and a second time when the decaying cyclones regain strength by incorporating fresh moisture over the eastern North Atlantic. Further case studies of such uptake sequences could help to clarify if this preliminary picture is indeed appropriate.

While it should again be stressed that the present work allows only to derive firm conclusions about the moisture that is transported to Greenland, the impression from Fig. 3.13 is that the pre-uptake history of moisture shows a robust picture that gradually changes with the NAO index. Hence, the identified pre-uptake history probably provides a general picture of the interrelations between dry-air outbreaks and evaporation in the North Atlantic during winter.

3.6 Further discussion and conclusions

The Lagrangian moisture source diagnostic developed in this chapter revealed new insights into the processes of moisture transport to Greenland. However, being a new method, there is a need to reconsider some of the assumptions, limitations and caveats, before the implications of these new findings are compared to previous views on the moisture sources for Greenland.

3.6.1 Method discussion

The present methodology is necessarily built on a number of assumptions, which, depending on the desired application, can be important to consider:

- The Lagrangian precipitation estimate is derived from a forward projection of the moisture decrease during the last 6 h before arrival onto the arrival location, which causes a high bias (see Sec. 3.1). An improved estimate could be derived by distributing the precipitation over the area crossed by the air parcel during the last 6 h before arrival. Remedy however appears limited, since due to orographic drag, air masses are slower over land than over sea. A second simplification is that all microphysical processes are neglected in our precipitation estimate, and any decrease in specific humidity is directly assumed to be due to precipitation.

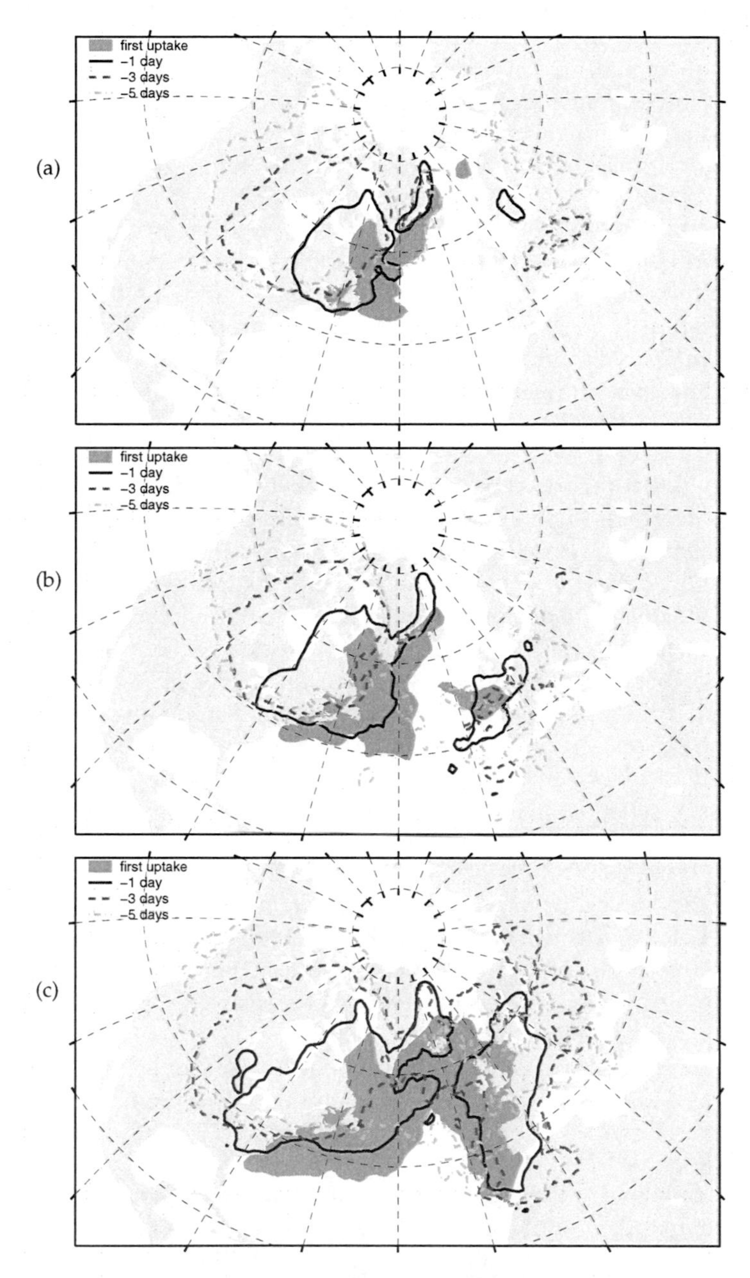

Figure 3.13: Mean pre-uptake locations during (a) NAO+, (b) NAO=, and (c) NAO− months. Shaded: first uptake location, contours: 1,3, and 5 days prior to first uptake. Contours are drawn at 20% of the maximum value of the respective frequency distribution.

This is quite different to the precipitation parametrisation in the ECMWF model, where condensating moisture has to take the pathway via the liquid and/or ice phase to precipitation. This necessarily leads to an overestimation of Lagrangian precipitation estimates (Stohl and James 2004). Hence, qualitatively, our precipitation estimate compares well with the ERA40 forecast precipitation, while quantitatively ~20-40% overestimation have to be taken into account.

- In order to diagnose moisture uptake in the boundary layer, the Lagrangian air parcel concept has to be relaxed so that their surface become permeable to turbulent influx of moisture. While this seemingly questions the Lagrangian concept of materially conserved air parcels, at the scale of air parcels considered here (60×60 km, 30 hPa), turbulent exchange will hardly decompose or replace the air parcel completely during a 6 h interval, and keeps the concept valid for the approach of this study.

- It is known that for trajectory calculations beyond 10 days, the coherency of an air mass decreases significantly, and wind field errors can lead to unacceptable deviations from the actual movement of air parcels (Stohl 1998). For the calculation time of 20 days, which was applied here, the influence of calculation errors can become unacceptably large. However, as was seen in Section 3.4, virtually all relevant moisture uptake and transport with respect to the precipitation at arrival takes place on shorter time scales (3–7 days), and hence the increasing uncertainty for trajectory lengths beyond 10 days does not affect the present results, nor are such long calculation times required to apply the moisture source diagnostic in the Atlantic and for Greenland.

- A number of moisture transport processes had to be neglected in the current Lagrangian framework. This includes moisture changes due to convection, turbulence, numerical diffusion, and rainwater evaporation, which all could lead to unattributable moisture increments above the BL. For winter-time moisture transport to Greenland, convective activity should be generally low, even though it may play a role over more southerly ocean areas. The identification of above-boundary layer moisture uptakes (Fig. B.2) is a first step towards closing this gap. In addition, using particle models, it may in the future be possible to consider convective water transport in a Lagrangian framework (Stohl and James 2005).

- While the net effect of moisture changes in an air parcel actually reflects P-E (Eq. 3.1), only one of the processes is assumed to dominate at a 6 h time interval. If so, negative Δq are precipitation events and positive Δq reflect evaporation, which is a plausible assumption (Stohl and James 2004). While the results found here also suggest validity of the P-E assumption, a rigorous check would require a comparison of the results with trajectory calculations at a shorter time interval (e.g. 1 h) and on a finer spatial scale, e.g. with RCM simulation data. Even more powerful could be a comparison with an Eulerian method for identifying water sources, since such a method would also take into account the moisture transport processes omitted here.

- Comparing this method with previous Lagrangian moisture source diagnostics, several similarities and differences can be noted. The moisture diagnostic of Dirmeyer and Brubaker (1999), also applied in studies by Brubaker et al. (2001) and Reale et al. (2001), has several problematic characteristics. First, their backward trajectories are calculated quasi-isentropically, which is a severe limitation when diabatic processes are important. Second, surface fluxes are allowed to contribute to moisture increase along trajectories overhead, even when these are far from the boundary layer. The Lagrangian moisture source identification of James et al. (2004), later extended to a particle model by Stohl and James (2004, 2005), has a good qualitative but overestimated quantitative agreement of the precipitation estimate with observations in common with the method developed here. The major difference between our approach and that of James et al. (2004) and Stohl and James (2004, 2005) is the introduction of a BL criterion and an attribution of moisture sources, which for the first time is applied here. The two innovations allow for a clearer determination of moisture sources than the previous approaches. Furthermore, the source attribution limits the amount of the identified moisture uptake to that of the precipitation estimate, and avoids the up to 7-fold overestimations found by Stohl and James (2004).

In summary, a robust Lagrangian moisture diagnostic has been developed and tested successfully. Nevertheless, a future comparison with other approaches could be very insightful.

3.6.2 Moisture transport analyis

With respect to the analysis of moisture source variability for winter-time precipitation in Greenland, new insight could be gained. At the same time, some limitations remain, and new questions have to be posed:

- The results found here only represent the moisture that is transported to and falls as precipitation over Greenland. The moisture sources found here are hence only a subset of the overall evaporation over the North Atlantic, which may show a quite different pattern.
- No conclusions on the seasonality of moisture transport to Greenland are currently possible, since this study is only concerned with winter precipitation. An analysis of the seasonal variability of these moisture sources could be a very promising future application of this method.
- The Lagrangian precipitation estimate for the Greenland plateau compares reasonably well with the ERA40 precipitation. The moisture source diagnostic allows

to attribute moisture sources to ~59% of the diagnosed precipitation. Including the above-boundary layer uptake of moisture, which to some degree could be representative of the larger surroundings, the fraction of (broadly) attributed precipitation increases to ~84%.

- The moisture sources identified from this methodology are clearly different than found in some earlier studies. The fixed subtropical moisture source in the western North Atlantic assumed by Johnsen et al. (1989) and Barlow et al. (1993) clearly seems inappropriate. The distribution of moisture sources identified in two GCM tagging studies (Charles et al. 1994; Werner et al. 2001) agrees reasonably with the results found here. A new aspect of this study is that a strong variability of moisture sources with the NAO was identified, a fact which was previously not known. Moisture sources for higher (lower) elevations of the Greenland ice sheet have more southerly (northerly) moisture sources. Inter-annual variability is non-uniform for the southern and the central to northern parts of Greenland.
- Using a moisture source attribution leads to a strong shift of the maxima of moisture uptake towards the arrival region. Uptakes in further remote source areas, such as the Pacific and the Gulf of Mexico, are strongly discounted. The reduced importance of remote sources is certainly in line with expectations from time scales of tropospheric water transport. This highlights the importance of using a source attribution algorithm for Lagrangian moisture source diagnostics.
- Two case studies suggest that during winter, land-sea contrasts and anticyclones play a fundamental role for the evaporation of moisture in the North Atlantic, which is then transported with mid-latitude cyclones to Greenland. In future studies with the methodology developed here, it will be possible to gain new insight into the interaction and importance of these processes for atmospheric moisture transport.

The newly identified moisture sources have considerable implications for the understanding of Greenland isotopic records. Hence, in the next chapter, these results will be exploited for an investigation of the NAO influence on the isotopic signal in precipitation over Greenland.

Chapter 4

Lagrangian Isotope Modelling

This chapter combines the results from the Lagrangian moisture source diagnostic (Chapter 3) with the isotopic concepts described in Chapter 2 to model the NAO variability of the isotopic signal in Greenland precipitation for present-day winter conditions. Two features of the present study should be highlighted: (i) The moisture sources diagnosed with our new method are used to determine the moisture's evaporation conditions. (ii) As this study is based on reanalysis data, a direct comparison with present-day stable isotope observations from Greenland is possible.

After introducing the parameters which are important for isotopic fractionation on the way from the moisture sources to Greenland, the diagnosed isotopic fractionation conditions are presented in the sequence of moisture transport in an air parcel: First for the evaporation, then at condensation onset, and finally for arrival over Greenland. After a sensitivity experiment with the fractionation model and a preliminary comparison of isotopic fractionation modelling results to selected site observations, a comprehensive discussion concludes this chapter.

4.1 Isotopic fractionation parameters

As was laid out in Chapter 2, isotopic fractionation in the atmosphere is strongly dependent on temperature. In a Lagrangian framework, different temperatures and temperature ranges (among several other parameters) are important at different stages of the transport sequence of moisture in an air parcel (Fig. 4.1). The concept developed in the previous chapter is used to diagnose these fractionation parameters along the same backward trajectories.

During evaporation from the sea surface (Fig. 4.1, ①), according to the Craig-Gordon model (Craig and Gordon 1965), isotopic fractionation depends on the sea surface temperature (SST), relative humidity (RH), 2 m air temperature in the boundary layer (T_{2m}), and wind velocity (U_{10m}) above the surface. These parameters are diagnosed as space-time averages from the ERA40 reanalysis data as described in Sec. 3.1.

Boundary layer (BL) air with a given isotopic composition will eventually be turbulently mixed into a bypassing air parcel.

As the air parcel moves away from the BL into the free atmosphere, adiabatic and radiative cooling will lead to condensation and cloud formation. The occurrence of condensation in an air parcel denotes the onset of equilibrium fractionation between water vapour and cloud water (Fig. 4.1, ②). Condensation temperature T_{cnd} is diagnosed here at the first instance where clouds form in an air parcel (RH≥80%, c.f. Sec. 3.1) after the last uptake and before arrival in Greenland. Note that T_{cnd} is diagnosed only once along each trajectory.

Isotopic fractionation continues in the air parcel during transport to Greenland with continuing condensation due to adiabatic and radiative cooling and precipitation, until the minimum saturation vapour pressure is reached (Fig. 4.1, ④). In all except very few cases ($< 0.01\%$), the arrival location of the air parcels over Greenland was also the location of minimum vapour pressure (not shown). Therefore, the temperature at the arrival location and altitude of an air parcel (T_{arr}) is diagnosed as the end point of isotopic fractionation. Condensed moisture at the arrival location falls as precipitation to the surface. In the case of solid precipitation (snow) and a sufficiently low melting layer position, which both are generally the case for Greenland winter conditions, the isotopic values are transferred without changes to the surface (Sec. 2.2). Therefore, T_{arr} should at least to some degree be reflected in the isotopic composition of precipitation at the surface. Krinner et al. (1997) and Krinner and Werner (2003) accordingly used a precipitation-weighted arrival temperature (*precipitation temperature*) as a proxy for isotopic fractionation effects in general circulation models (GCMs) without isotope tracers.

In addition to the absolute temperature values of condensation onset and end which influence the fractionation factors α (see Sec. 2.1.2), the temperature range between T_{cnd} and T_{arr} is important, as it determines the extent of isotopic equilibrium fractionation (Dansgaard 1964; Jouzel et al. 1997). This temperature range is therefore defined here as the temperature difference T_{dif}, later in this chapter also termed *fractionation range*:

$$T_{dif} = T_{arr} - T_{cnd}. \tag{4.1}$$

At the site below the arrival of the air parcel, where precipitation is deposited, two further temperatures are diagnosed (Fig. 4.1, ④). The skin temperature SKT[1] and the air temperature 2 m above the ice shield surface (T_{sfc}) are obviously not related to isotopic fractionation itself, but reflect the surface conditions at the precipitation site. As they are part of the empirical isotope-temperature relationship (see Sec. 2.5), these temperatures are also diagnosed at the arrival location for each precipitating trajectory.

[1]calculated from radiative energy balance closure at the surface

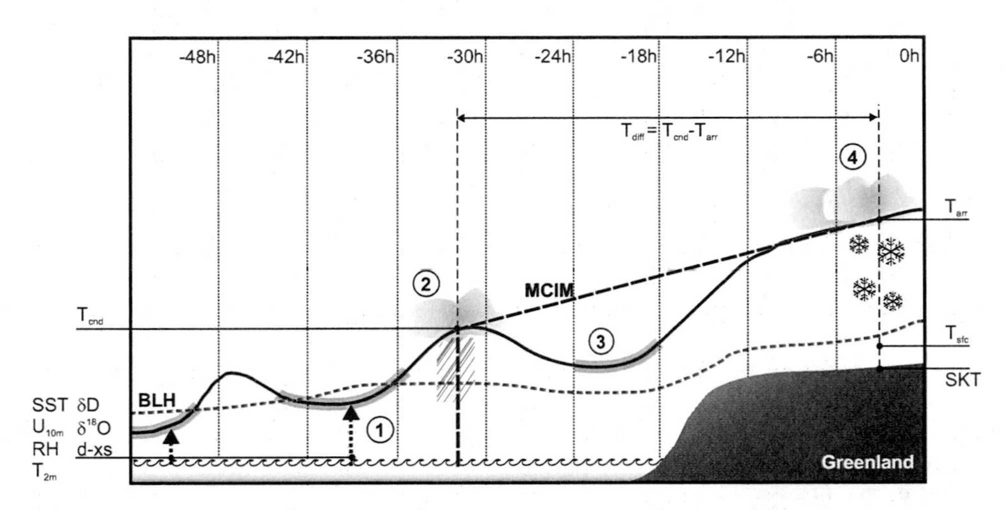

Figure 4.1: Sketch of the definition of temperatures affecting the isotopic fractionation process along a backward trajectory (blue line). Top row: time before arrival. BLH: boundary layer height. Thickened sections along the trajectory denote sections of moisture increase. Dashed orange line shows the idealised fractionation trajectory of the MCIM model. See text for details.

4.1.1 Isotopic fractionation from diagnosed parameters

The temperatures and fractionation parameters diagnosed along the transport path of moisture serve as input for the MCIM[2] stable isotope model (Ciais and Jouzel 1994, see Appendix C for a more detailed description of the model). MCIM is a Rayleigh-type isotope model with parametrised mixed-phase microphysical processes, which calculates fractionation along an idealised $T - p$ trajectory. In Fig. 4.1, the MCIM fractionation trajectory, which corresponds to the actual air parcel trajectory (blue) is indicated by the dashed orange line. Originally, MCIM calculates the condensation temperature from adiabatically lifting an air parcel from the sea surface to its lifting condensation level (LCL). For this study, in order to simulate fractionation over the temperature range from T_{cnd} to T_{arr} for each air parcel, a modified lapse rate and surface RH were calculated which lead to an LCL that coincides with the diagnosed condensation temperature and pressure (see Appendix C). Otherwise, the same setup as described in Masson-Delmotte et al. (2005a) was applied. Note that this implies a tuning of model-internal parameters which represent microphysical processes to moisture source temperatures of 15–20°C. Isotope fractionation is calculated separately for each moisture uptake event with the modified MCIM model, and the isotopic composition of the precipitation at the arrival site is composed by each source's contribution (fraction f, see Sec. 3.2) to the arrival precipitation.

[2]Mixed-Cloud Isotope Model

4.1.2 Initialisation with GCM isotope data

A problem of this model application concerns the initial isotopic composition of the water vapour in the BL. As has been shown by Jouzel and Koster (1996), the approach to calculate the isotopic composition of the evaporating sea water with a global closure equation, such as in the original MCIM model, is inconsistent with observations. They recommended instead to use the stable isotope composition at the lowest layer of isotope GCMs as initial values for Lagrangian isotope models. Therefore, in addition to the evaporation parameters at the moisture source, the isotopic composition (δD, δ^{18}O) of the water vapour in the lowest model layer of the ECHAM4 isotope GCM (Hoffmann et al. 1998; Werner et al. 2001) was extracted at the diagnosed moisture source regions. Gridded $2^\circ \times 2^\circ$ monthly mean data were acquired from the SWING database [3] in November 2005 for the same months as in the moisture source diagnostic (Table 3.2). The model was run for a 134-year period at T159L30 resolution, and driven by the Hadley SST data (experiment S1B). Deuterium excess was calculated from the monthly mean fields according to

$$d = \delta\mathrm{D} - 8 \cdot \delta^{18}\mathrm{O}. \tag{4.2}$$

As the isotope GCM simulations were driven by the Hadley SST dataset, the (observationally unconstrained) atmospheric data may be partly inconsistent with the ERA40 reanalysis data used here. However, since monthly mean files were used, and the surface isotope conditions should be dominated by the slowly varying SST, isotopic composition at the surface probably has only small day-to-day variability due to atmospheric influences.

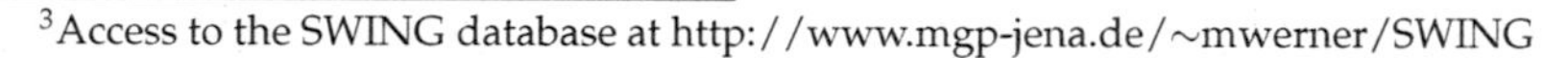

[3] Access to the SWING database at http://www.mgp-jena.de/~mwerner/SWING

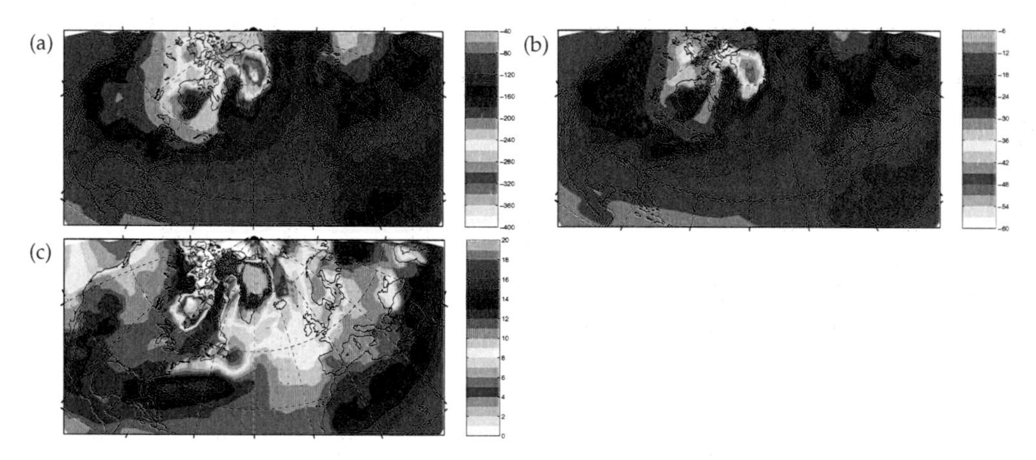

Figure 4.2: Mean distribution of the stable isotope parameters (a) δD (‰), (b) δ^{18}O (‰), and (c) d-excess (‰) at the lowest layer of the ECHAM4 isotope model for the mean NAO negative phase. Data are from monthly mean files of the SWING database.

Fig. 4.2 shows the stable isotope composition of water vapour at the lowest model level for the NAO negative (NAO−) phase, averaged over the 10 selected months. Values range between -40 to -120‰ for δD and -6 to -40‰ for δ^{18}O. These values are twice as depleted as from linear equilibrium fractionation considerations alone (see Sec. 2.2), which demonstrates the effect of the isotope GCM fractionation parametrisations. Both, δD and δ^{18}O exhibit a mostly zonal gradient across the North Atlantic, in agreement with SST fields (c.f. Fig. 3.10). The gradient weakens towards the eastern North Atlantic. The d-excess (Fig. 4.2c) shows a strong gradient near the area of the strongest SST gradient (c.f. Fig. 3.10e, f). Minimum and maximum values are 5‰ at 50°N and 15‰ at 30°N, respectively. In other areas, d-excess is ~8–10‰. It is noteworthy that both the surface water vapour isotope fields and the SLP field of the ECHAM4 simulation (not shown) show almost no NAO variability for the selected periods.

In order to test the influence of the different initial conditions, two isotopic fractionation experiments are performed: A first one using the isotope GCM surface vapour composition, and a second one using the original MCIM with the global closure of Merlivat and Jouzel (1979).

4.2 Moisture uptake conditions

According to the Craig-Gordon model (see Sec. 2.2) that describes fractionation during evaporation from a water body, equilibrium fractionation occurs in the immediate transition from the sea surface to the saturated layer directly above. Then, non-equilibrium fractionation occurs under the influence of an RH gradient in a transition layer that is dominated by molecular diffusion. In the turbulent BL above, transport takes place without further fractionation. Our Lagrangian methodology detects moisture uptake of an air parcel in the BL rather than instantaneous evaporation. While this is not exactly the same, it is assumed here that the horizontal transport in the BL is negligible for the air parcels' dimensions considered, and the uptake conditions are largely similar to the evaporation conditions.

4.2.1 Diagnosed evaporation conditions

The diagnosed evaporation conditions (SST, T_{2m}, RH, and U_{10m}) are now examined as Lagrangian forward projections onto the respective arrival location on the Greenland plateau, weighted by their contribution to the local grid-point precipitation (Fig. 4.3).

For the NAO+ phase, the precipitation at coastal areas evaporated at SST of 4–7°C for coastal areas in Greenland, while the interior of the plateau is associated with source region SSTs of up to 10–12°C (Fig. 4.3a). The changes for the NAO− phase are remarkable: A SW-NE oriented band of higher source region SSTs (11–12°C) is apparent, while values are lower at the slopes (7–8°C). Hence, the north-east of the plateau undergoes source region SST changes of ~8–10 K with NAO phase, while most other areas experience 4–5 K SST shift. The overall shift to warmer source region SSTs is also apparent in the SST histogram. The pattern of source SST changes strongly resembles the distribution of source latitudes over the Greenland ice sheet (Fig. 3.9b). This shows that NAO-induced circulation changes control the SST variability of the moisture sources for Greenland, rather than local SST variability at a fixed moisture source.

This view is partly in disagreement with previous studies. Using an early conceptual model, Johnsen et al. (1989) concluded that moisture sources for Greenland must be tropical, as their fractionation results showed best agreement with a source latitude of 35–40°N and source region SSTs of ~22–26°C. Barlow et al. (1993) interpreted changes in δD and d-excess with NAO as SST variability at a fixed moisture source at that same latitude. While White et al. (1997) still assumed a dominant moisture origin of 20–30°N for Greenland, Barlow et al. (1997) argued for more local moisture sources. The combined isotope/tagging GCM study by Werner et al. (2001) presented evidence that different moisture sources with differing isotopic depletion contribute to the isotopic precipitation signal in Greenland. Masson-Delmotte et al. (2005a,b) recognised that moisture source changes could also lead to differences in the ice-core signals at different drilling sites, rather than varying SST. This latter view is supported by the findings of this work.

The 2 m-temperature at the evaporation locations (T_{2m}) has a pattern very similar to the SST, yet with an almost uniform negative offset of 3–4 K (not shown). A slight

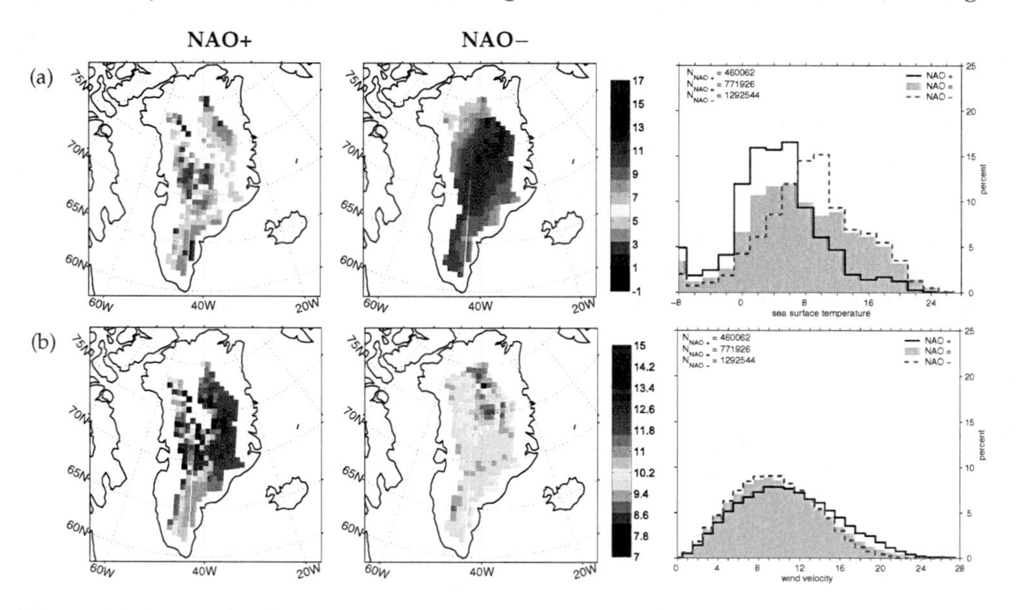

Figure 4.3: Lagrangian forward projections and histograms of the evaporation conditions at the moisture uptake locations diagnosed from the ERA40 reanalysis data. (a) SST (°C), (b) U_{10m} (m s^{-1}).

tendency towards larger differences between SST and T_{2m} towards colder temperatures can be noted. This temperature difference is further examined in Sec. 4.5.

Wind velocities at 10 m (U_{10m}) for the NAO+ phase are rather high along the east coast of Greenland (10–15 $m\,s^{-1}$), and lower at the west coast (Fig. 4.3b). For the NAO− phase, wind velocities at the evaporation location are more uniformly distributed and on average lower (9–11 $m\,s^{-1}$). The pattern of U_{10m} is similar to the source longitudes (Fig. 3.9a), but is probably a result of the stronger pressure gradient in NAO+ months. This pronounced difference was unexpected, and warrants further investigation, since higher surface winds can be associated with a different kinetic fractionation regime due to sea-spray evaporation (J. Gat, *pers. comm.*, 2006). In general, wind velocities are relatively high, which could reflect that the Lagrangian method preferentially diagnoses moisture uptake when air parcels enter the boundary layer during strong wind conditions. Evaporation of water from the ocean surface into the marine BL may equally well have occurred before under calmer conditions. For the calculation of the initial isotopic composition of the water vapour in the MCIM model from the global closure equation (experiment 2), wind velocities are limited to 10 $m\,s^{-1}$, hence U_{10m} will not have any effects on the modelling results in that sensitivity experiment. Relative humidity at the evaporation locations is rather uniformly distributed at ~72% for the NAO+, and ~70% for the NAO− phase mean, and are hence not further discussed here (not shown).

4.2.2 Diagnosed isotopic composition

The surface vapour's isotopic composition is extracted from the lowest model layer in the SWING dataset at the same locations as for the uptake conditions. The extracted δD and $\delta^{18}O$ at the surface have patterns similar to the SST pattern (Fig. 4.4a, b). This demonstrates the control of SST on the isotopic composition close to the surface. For the NAO+ (NAO−) phase mean, values range between -130 to $-144‰$ (-112 to $-130‰$) for δD, and -17.0 to $-18.8‰$ (-15.2 to $-17.0‰$) for $\delta^{18}O$.

The clear NAO variability which is present in the isotopic composition of the initial water vapour highlights again the strong influence of the source region changes. Only small local NAO variability is visible in the water vapour's isotopic composition from the ECHAM4 simulations, hence the differences in initial vapour compositions for δD and $\delta^{18}O$ of on average 14‰ and 1.8‰, respectively, are almost exclusively due shifts in source regions.

The parameter d-excess shows a pattern that seems not directly related to SST (Fig. 4.4c). Values for d-excess range mostly between 8–10‰. For the NAO+ phase, more variability is visible in d-excess, probably due to the more westerly source regions (c.f. Fig. 4.2c). This could be an indication of the fact that d-excess is more dependent on the local RH than the SST conditions (Merlivat and Jouzel 1979; Jouzel and Koster 1996).

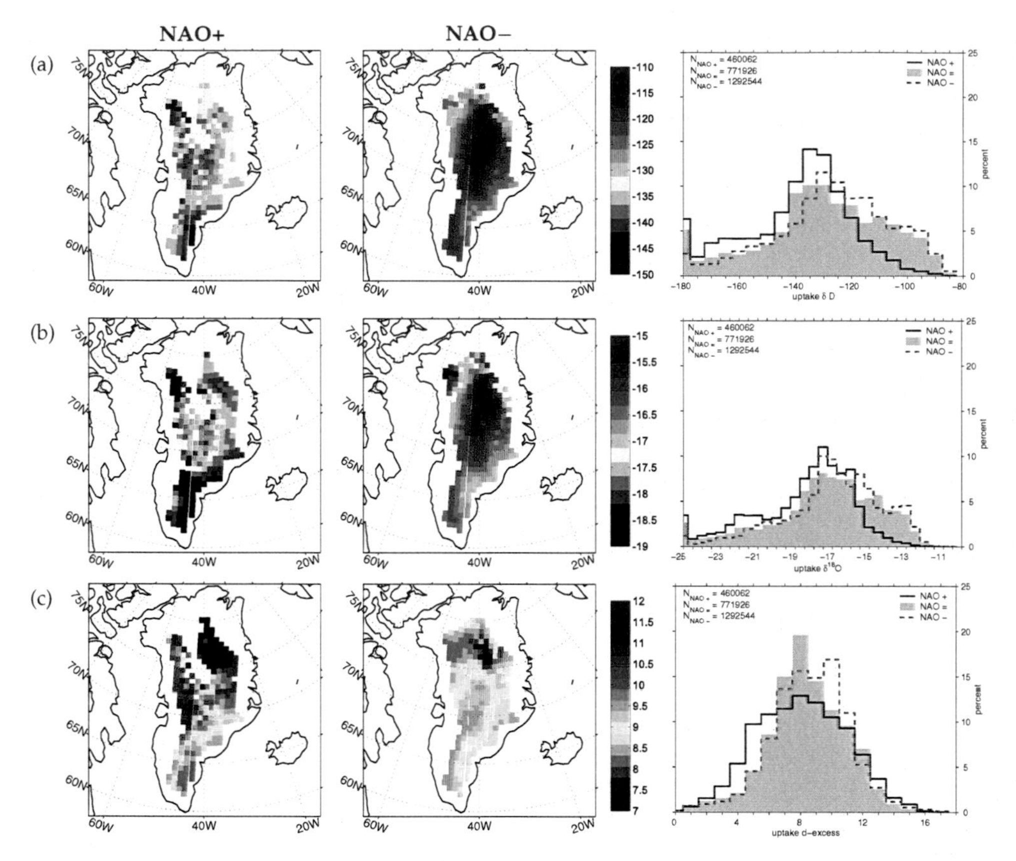

Figure 4.4: Isotopic composition of water vapour at the evaporation site at the lowest model level of ECHAM4 isotope GCM simulations from the SWING dataset: (a) δD (‰), (b) δ^{18}O (‰), (c) d-excess (‰).

4.3 Condensation onset conditions

Condensation onset locations are identified along each trajectory as the first occurrence of cloud formation (RH$\geq$80%) after the last moisture uptake before arrival over Greenland (Sec. 4.1). Figure 4.5 shows the spatial distribution of these diagnosed condensation onset locations for the three NAO phase averages. All three phases show a maximum of condensation onset at the south-eastern slope of Greenland. Condensation locations are more localised for the NAO+ phase, which corresponds to the rather confined moisture source regions (c.f. Fig.3.7). During NAO− months, condensation occurs over a considerably broader area (Fig. 4.5c), corresponding to the more south-easterly source regions during this phase. In contrast to the NAO+ phase, no condensation occurs over the Norwegian sea during the NAO− phase. The NAO= phase again shows a transition between the other two phases (Fig. 4.5b).

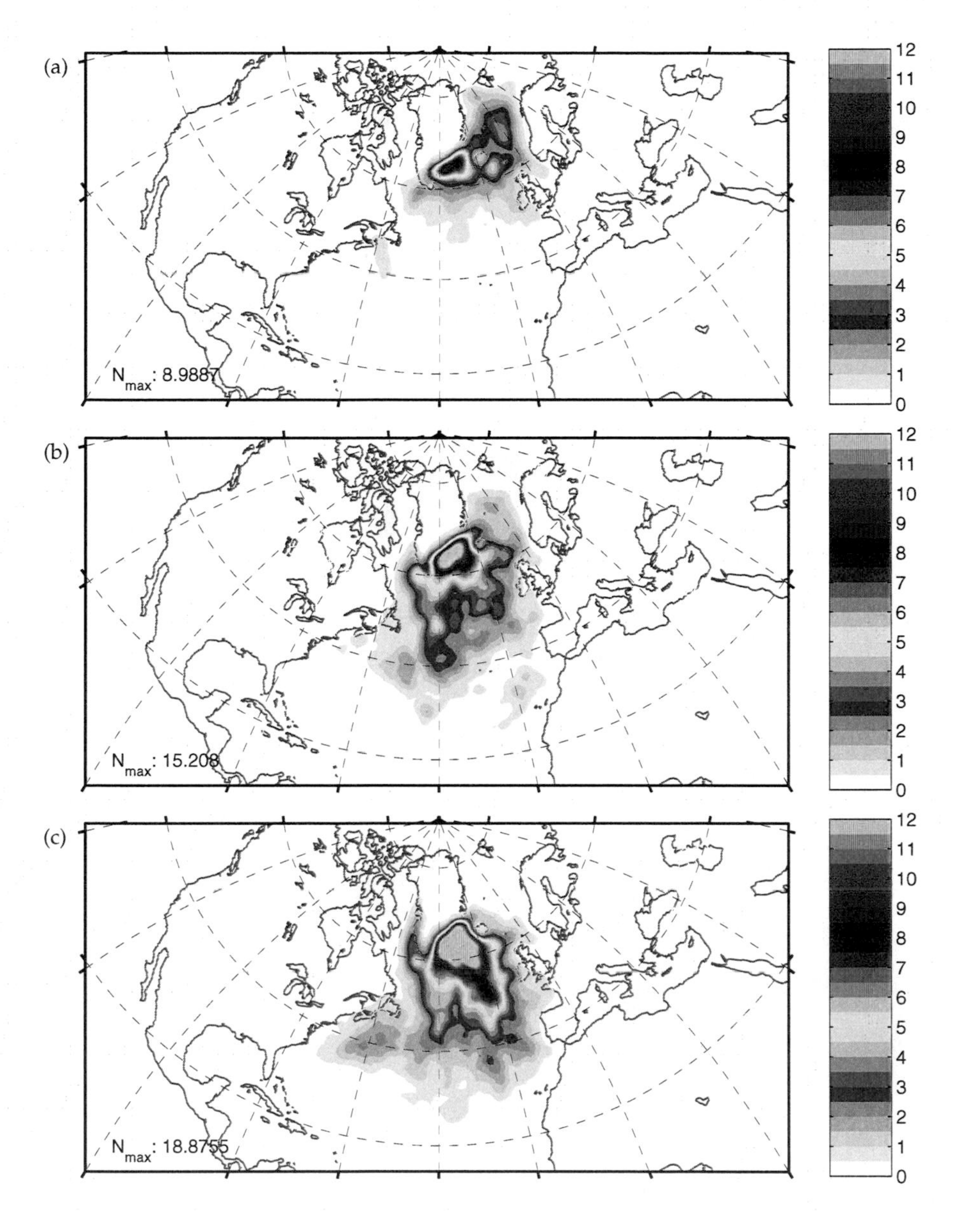

Figure 4.5: Phase mean condensation locations for (a) NAO+, (b) NAO=, (c) NAO− months, shown as contribution to the monthly mean precipitation in Greenland. Units are mm precipitation contribution integrated over 10^4 km^2.

Values of T_{cnd} range between -3 to $+3$°C and $+3$ to $+6$°C for the NAO+ and NAO− phase, respectively (Fig. 4.6a). The corresponding histogram shows a strong influence of the NAO: The phase T_{cnd} mean shifts from -1.4°C for NAO+ to 4.5°C for NAO−

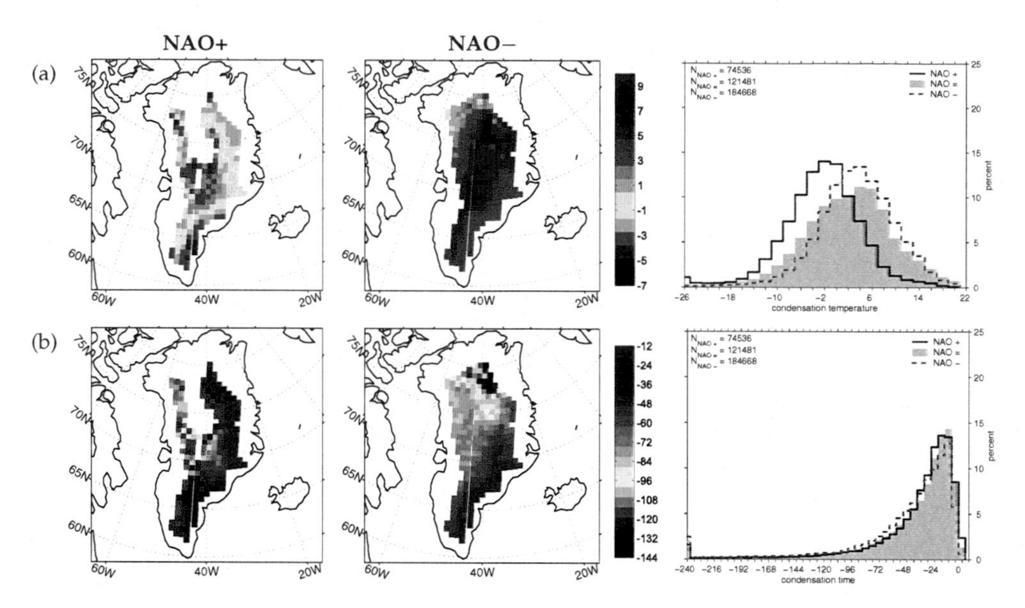

Figure 4.6: Lagrangian forward projections and histograms of the conditions at the diagnosed condensation onset locations. (a) Condensation temperature T_{cnd} (°C), (b) condensation duration t_{cnd} (h).

($\Delta T_{cnd} = 5.9\,K$). The regional patterns of the forward-projected T_{cnd} are rather similar to those of the SST (Fig. 4.3a), which suggests a similar influence of latitudinal transport.

The condensation duration t_{cnd}, which is the same as the time of condensation onset before arrival, is on average 11 h shorter for the NAO+ months ($-49.8\,h$) than for NAO− months ($-60.3\,h$, Fig. 4.6b). This difference is however not uniformly distributed over the plateau: While for the NAO+ phase condensation times are similar along the east coast, during NAO− a N-S gradient is evident, which is reminiscent of the transport time analysis (c.f. Fig. 3.9c). The earlier condensation time for the northern sector means that fractionation can take place over a longer period of time than for the more southerly and coastal regions. The more meridional transport during NAO− months is most likely the cause for this significant gradient in condensation time across Greenland.

The average condensation pressure p_{cnd} (not shown) is diagnosed as $\sim$930 hPa, which corresponds to an altitude of $\sim$ 1000 m asl. NAO variability of the condensation pressure is small, but still on average 18 hPa lower in the NAO+ than in the NAO− phase. This could reflect the lower mean pressure during the NAO+ phase near the condensation locations due to the more confined storm track (comp. Fig. 3.10a).

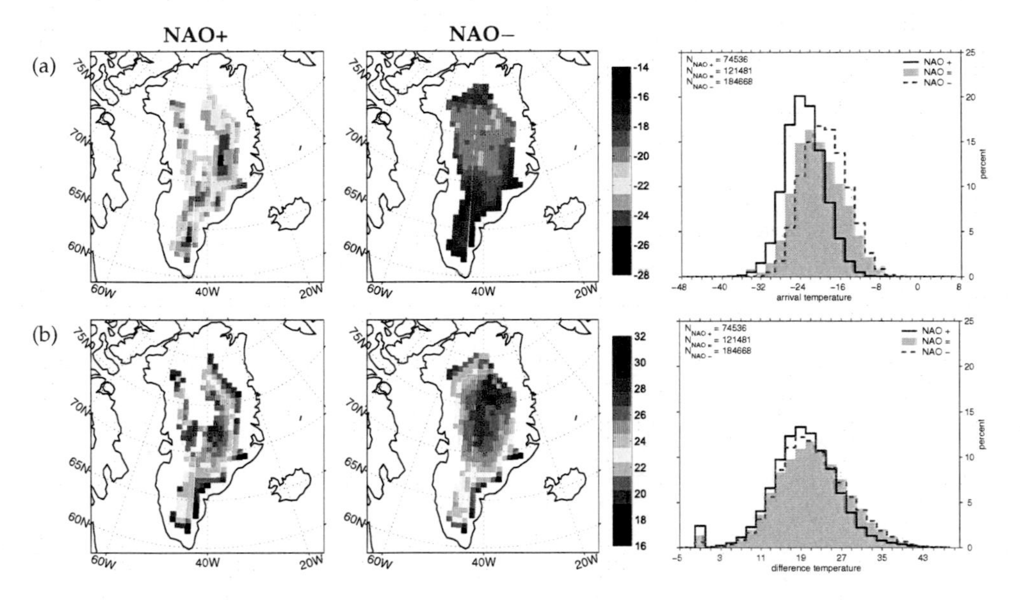

Figure 4.7: Lagrangian forward projections and histograms of moisture arrival conditions, weighted by precipitation amount. (a) arrival temperature T_{arr} (°C), (b) temperature difference T_{dif} (K).

4.4 Arrival conditions

Temperature and pressure at the altitude of the arriving moisture over Greenland determine the end point of isotopic fractionation. Figure 4.7a shows a strong NAO dependence of the arrival temperature T_{arr}. While during NAO+ months values are uniformly close to -24°C, T_{arr} shifts to a N-S-N pattern during the NAO− phase, with on average 4.2 K warmer conditions. The cold arrival temperatures during the NAO+ phase correspond to the temperature minimum in the mid-troposphere with a centre over eastern Greenland during that phase (c.f. Fig. 3.10c). Similarly, the warmer conditions during NAO− months correspond to warmer mid-tropospheric temperatures, with the exception of an anomaly in the very north of the plateau. The average arrival pressure of water vapour (~660 hPa) is lower over higher regions of the ice sheet (not shown), and almost invariant with NAO phase (Fig. 3.6a).

The temperature difference T_{dif} (Eq. 4.1) reflects the temperature range over which fractionation can take place (Fig. 4.7b). The interior of the ice sheet exhibits a larger T_{dif} than its borders (28 K vs. 20 K). Interestingly, despite the very different evaporation, condensation, and arrival conditions, the range and regional distribution of T_{dif} are quite similar for the two NAO phases. As a small difference between the two phases a shift of the maximum of T_{dif} towards the NE during NAO− months can be noted. This NAO invariance could suggest that the changes in moisture source regions and mid-tropospheric temperatures over Greenland cancel each other w.r.t. T_{dif}, at least to some extent.

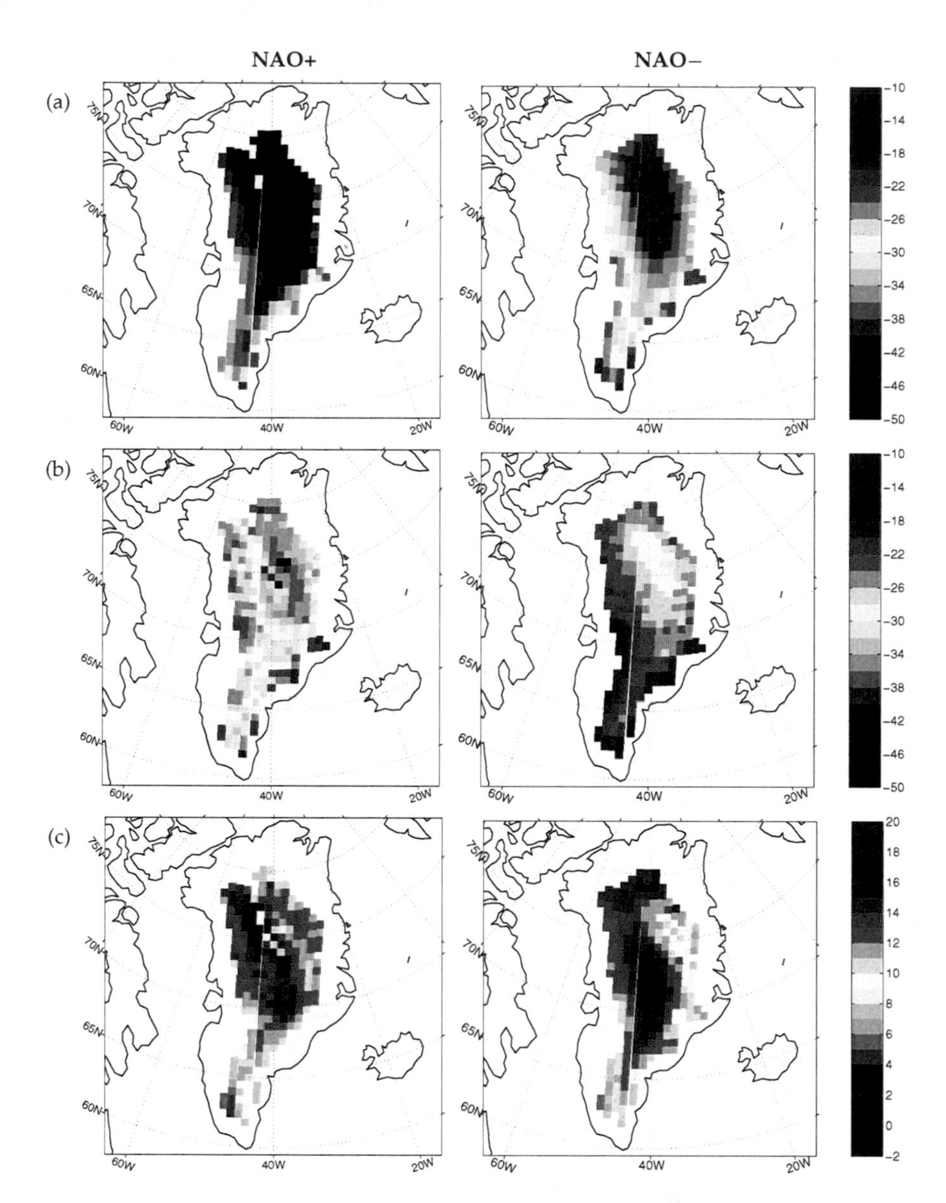

Figure 4.8: Mean skin temperatures over Greenland for (a) the 10-monthly mean (°C), (b) on precipitation days only (°C). (c) Skin temperature difference between precipitation days and the monthly mean (K).

4.4.1 Wet and dry condition temperatures

In comparison to typical annual mean station records of surface temperatures on the ice sheet, (e.g. $-32°C$ in the annual mean at Summit, Greenland, (Johnsen et al. 1989)), the arrival temperatures of moisture above Greenland found here may seem rather warm for winter conditions. Our moisture transport diagnostic records however only the temperatures during the arrival of precipitation, which may be very different from the longer-term mean. It is well known that high latitude regions can become significantly warmer than normal when mid-latitude disturbances advect moist and warm air, and change the radiation balance due to increased cloud cover (Loewe 1936), and remove the strong surface inversion that builds up under clear-sky conditions. The precipitation temperatures of Krinner et al. (1997) and Krinner and Werner (2003) are also in a similar range as found here.

To quantify the difference of our diagnosed arrival and surface temperatures to the winter mean conditions, ERA40 skin temperatures (SKT)[4] over the Greenland plateau are compared for dry and wet periods during the selected cases (Fig. 4.8). The mean SKT of all days of the 10 selected NAO+ (NAO−) months reaches below $-48°C$ ($-38°C$) (Fig. 4.8a). For precipitation periods only, SKT is significantly warmer (Fig. 4.8b). Over the highest region of the plateau, this difference is up to 16 K. The warm bias is very similar for the two NAO phase means, except for the anomalously warm northern regions during NAO− months (c.f. Fig. 4.7a). This could indicate that the precipitation regime during NAO− months induces the temperature anomalies in this area (Masson-Delmotte et al. 2005a). The difference between wet and dry temperatures is also of great importance for the isotope-temperature relationship (see Sec. 4.7.1).

4.5 Isotopic fractionation modelling

As introduced in Section 4.1.1, the MCIM model was applied for calculating isotopic fractionation of water vapour on the way from its source area to Greenland according to diagnosed temperature parameters. Two experiments were conducted: experiment 1, which is based on GCM initialisation conditions, and experiment 2, which is based on a global closure equation, and serves as a test of MCIM's sensitivity to the initial conditions. In both experiments, an adjusted lapse rate and surface RH are supplied as additional input parameters (see Appendix C).

4.5.1 Initialisation with GCM output

For experiment 1, the patterns of δD and $\delta^{18}O$ calculated from the GCM initialisation are very similar (Fig. 4.9a, b). Isotope ratios range between -220 and $-160‰$ for δD

[4]note that skin temperatures in high latitudes are more sensitive to radiative effects such as cloud cover than air temperatures (e.g. Sodemann and Foken 2005)

and between -28 and -20‰ for δ^{18}O during the NAO+ phase. For the NAO− months, ratios are between -200 and -120‰ for δD and between -24 to -16‰ for δ^{18}O. Hence, the mean of the frequency distribution shifts by 27.9‰ for δD and 3.4‰ for δ^{18}O, respectively, from NAO+ to NAO− months. The pattern of the initial isotopic composition is completely overprinted by the fractionation during transport (c.f. Fig. 4.4a, b). The histogram is also changed remarkably towards a more Gaussian-shaped distribution. The isotopic gradient for δD (δ^{18}O) from the central to southern plateau (75°N to 62°N) is ∼40‰ (∼5‰) during NAO− months. The pattern closely resembles that of the arrival temperature, which suggests a strong imprint of the temperature regime during fractionation, in particular the arrival temperature, on the final isotopic composition. If so, the NAO variability of stable isotope ratios in Greenland at least partly reflects the corresponding Northern Hemisphere temperature anomalies.

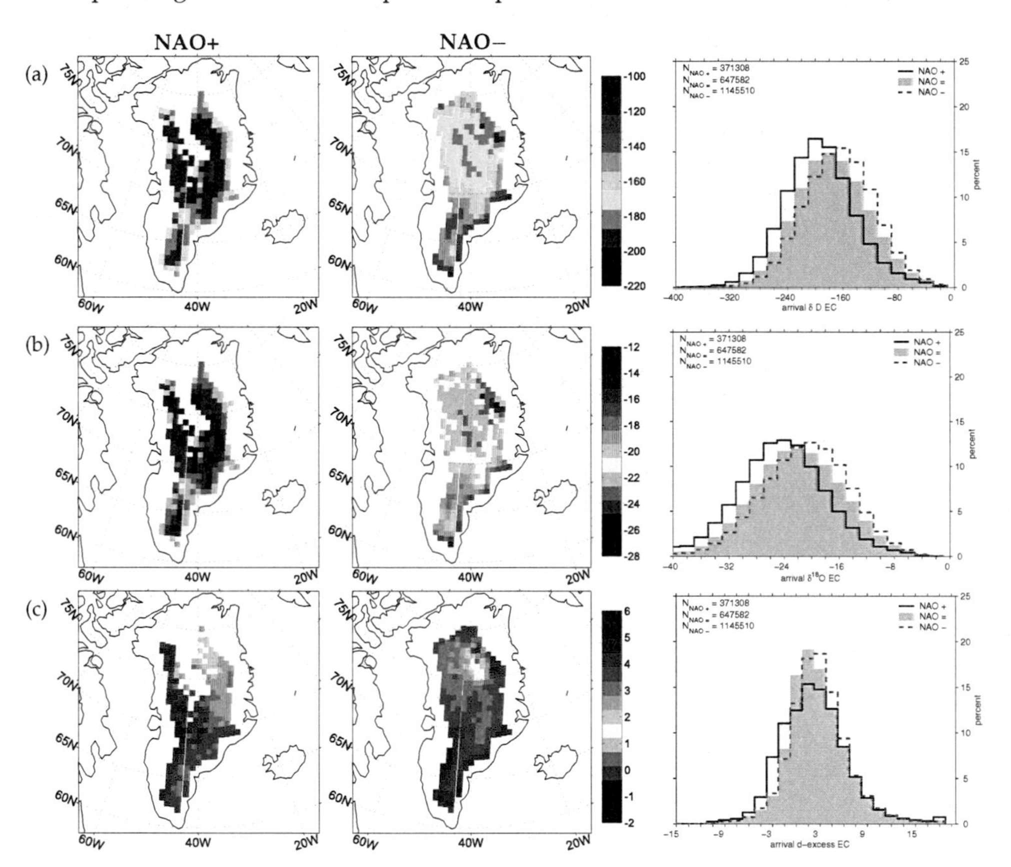

Figure 4.9: Isotopic composition of Greenland winter precipitation, calculated with initial conditions from the ECHAM4 isotope GCM (experiment 1). (a) δD (‰), (b) δ^{18}O (‰), (c) d-excess (‰).

The d-excess during NAO+ months ranges on average between 1–6‰, and slightly decreases from W to E (Fig. 4.9c). During NAO− months, a similar range with a more S-N oriented distribution is apparent. The mean shifts only by 0.5‰, but regional changes can be as large as 3‰ (e.g. in the NE of the plateau). Interestingly, d-excess increases with NAO index along the western edge of the plateau, while the reverse applies on the eastern fringe. The simulated d-excess on the plateau shows no obvious trace of the initialisation pattern (c.f. Fig. 4.4c). Since during transport d-excess is most strongly influenced by a parametrisation of kinetic fractionation effects related to ice supersaturation in mixed-phase clouds (see Appendix C), this change probably derives from there.

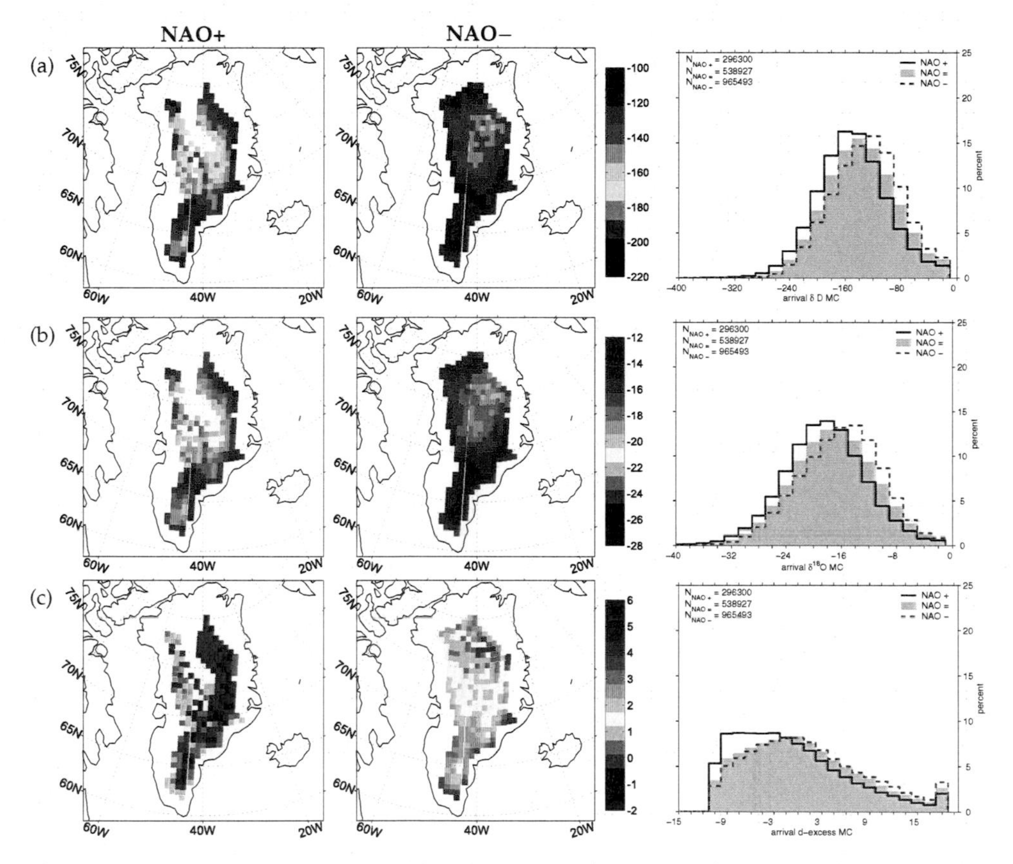

Figure 4.10: Isotopic composition of Greenland winter precipitation, calculated with initial conditions from a global closure equation (experiment 2). (a) δD (‰), (b) δ^{18}O (‰), (c) d-excess (‰).

4.5.2 Initialisation with surface parameters

In this sensitivity experiment (experiment 2), δD and δ^{18}O are clearly shifted towards less depleted values in comparison with experiment 1 (Fig. 4.10a, b). δD and δ^{18}O range only between -190 to -100‰and -22 to -14‰, respectively. This pronounced shift towards less depleted isotope ratios can be fully ascribed to the known deficiencies in the calculation of the initial vapour composition. The NAO variability is reduced by about 30% in comparison to experiment 1 (mean shift of 20.2‰ for δD and 2.3‰ for δ^{18}O from NAO+ to NAO−). This indicates that the isotopic composition of the initial moisture has a discernible influence on the final values. Therefore, the variability of the initial conditions calculated with the global closure equation from the diagnosed surface conditions must be smaller than in the GCM initialisation. A direct comparison of the water vapour's initial isotopic composition calculated from the global closure to the GCM fields could give further useful insight into MCIM's sensitivity.

The d-excess is considerably lower in this experiment than in the previous experiment (Fig. 4.10c). Values are on average close to 0‰, and show a strong NAO variability (0.0‰ for NAO+ and 2.1‰ for NAO−). The corresponding histogram shows that the calculation of d-excess is probably limited to values larger than -10‰ by an internal threshold in MCIM. This points again to severe problems in parametrising kinetic fractionation during evaporation with the diagnosed surface conditions (Jouzel and Koster 1996).

4.5.3 Comparison with observational data

A preliminary attempt is now undertaken to tie the MCIM modelling results of experiment 1 to observational data of δ^{18}O from Greenland ice cores (B. Vinther, *pers. comm.*, 2005). The two main foci are hereby to examine (i) how the absolute values of the modelled isotope ratios compare with the observed range of values, and (ii) how the NAO variability compares between modelled and observed isotope data.

To be useful in this comparison, observational data should represent a winter rather than an annual mean signal. Ideally, several of the months studied here should be in one common winter season (DJF). These requirements were met for three periods, namely the winters of 1964/65, 1968/69, and 1983/84 (Table 4.1). For the first two winters, we used season mean values of δ^{18}O (B. Vinther, *pers. comm.*, 2005) from the Alfabet cores A, B, D, E, and G (70.63–71.76°N; 35.82–39.62°W, 3018–3138 m asl). These shallow cores were drilled in an inverse L-shaped array southeast of the drilling site Crête in central Greenland. Data for winter 1983/84 is not covered by the shallow cores and was therefore taken from snow pits. Winter season isotope signals were calculated from an average of the five cores, where winter was defined as 35% of the annual accumulation, centered around the lowest δ^{18}O value (Vinther et al. 2003). There are significant differences between the five cores, as signal to noise ratios of single core seasonal δ^{18}O values are about 0.6 for central Greenland ice cores (B. Vinther, *pers. comm.*, 2005). The ice-core data show that mean δ^{18}O covaries with the NAO index at this site. The range of the values for the selected winters is 3.4‰ (Table 4.1). Currently, no standard deviation is available for the three winter averages.

The observational data are compared with a monthly average of the modelling results for the 4 grid points which cover the drilling sites. Considerable month-to-month variability is present in the modelled mean isotopic composition. However, as we use reanalysis data in this study, the underlying atmospheric variability should also be reflected in the observations. To obtain a seasonal estimate, the grid-point averages are weighted by their monthly mean precipitation values (Table 4.1). The mean isotope composition for the winter of 1964/65 and 1968/69 are therefore dominated by the months 196502 and 196901, respectively. It should be emphasised here that this comparison has a rather preliminary character, and more detailed comparisons should be conducted to more firmly couple observations and isotope modelling results.

For all three winters the observed values of $\delta^{18}O$ are substantially lower (10–12‰) than the model results (Table 4.1). For ice cores from more southerly locations, the offset decreases to about 6–8‰ on average (B. Vinther, *pers. comm.*, 2005). This means that isotopic fractionation is considerably underestimated when using MCIM with the diagnosed fractionation conditions, even when water vapour is initialised from isotope GCM data. The possible reasons for this offset are discussed in Sec. 4.5.4.

Even though the modelled isotope values are not in good absolute correspondence with observations, the NAO variability at the Alfabet site is met to a surprisingly high degree, both qualitatively (more depletion during NAO+ winters) and quantitatively ($\Delta\delta^{18}O \approx 2$‰ on average). The average change in $\delta^{18}O$ with NAO for the whole ice sheet is in even closer agreement (3.4‰). This indicates that the factors which influence isotopic NAO variability over Greenland are indeed captured by the model calculations. In section 4.6 an attempt is undertaken to disentangle the relative influences of the NAO variability.

Table 4.1: Comparison of observed winter season stable isotope data from the Alfabet cores ($\delta^{18}O_{Obs}$) with MCIM isotope model results ($\delta^{18}O_{MCIM}$) for corresponding grid points. Model results are averaged over 4 grid points and weighted by the respective monthly mean precipitation (P mean). Currently, no standard deviation is available for the three winter averages.

Month	NAO index	P mean (mm)	$\delta^{18}O_{Obs}$ (‰)	$\delta^{18}O_{MCIM}$ (‰)
198312	0.4	6	-38.43	-26.0
198401	4.1	6		
196412	-0.2	12	-35.06	-24.1
196501	-0.3	12		
196502	-5.1	37		
196901	-4.3	13	-36.32	-23.9
196902	-4.0	6		

Currently, no δD data are available from the Alfabet cores, hence no direct comparison to the d-excess can be carried out. In a study based on data from the GISP2 core, Barlow et al. (1993) investigated the concurrence of extreme values of the NAO with extreme values of δD and d-excess. For an ice-core section spanning the years 1840–1970, d-excess was on average 9.4 ± 1.6 for NAO+ and 8.7 ± 2.0 for NAO− winters. Note that the d-excess from the MCIM model results is only about half as large, even though in the initial conditions it was $\sim$ 10‰ (c.f. Fig. 4.4). However, there is a lagged seasonal signal in ice core d-excess, with maxima in autumn and minima in spring, which, taken into account, could bring observations and model results in better agreement.

4.5.4 Assessment of the MCIM isotope model

While absolute isotope ratios are clearly underestimated by MCIM for the current model setup, and the results for d-excess are not conclusive, the NAO variability is simulated quite reasonably. The possible reasons for these findings and the implications for using the MCIM model are discussed in the following.

As detailed in Appendix C, the MCIM model was originally designed to simulate the realistic isotopic ratios for a moisture source temperature of 20°C and a surface annual mean temperature at the arrival site of -30°C. In the context of this work, MCIM is applied to diagnosed rather than prescribed moisture transport and isotopic fractionation conditions. As these conditions strongly deviate from the original parameter choice and hence the tuning of the internal microphysical parameters, the discrepancies between the observed and simulated levels of δ^{18}O are not surprising. However, it should be noted that Petit et al. (1991) already found an offset of $\sim$10‰ when modelling the isotopic composition of Antarctic snow with an older version of the MCIM model.

An additional influence which could lead to an underestimation of the isotope depletion is related to the different fractionation conditions along idealised and calculated trajectories (c.f. Fig. 4.1). In contrast to an idealised fractionation trajectory, 'real' trajectories can include several cycles of condensation, which lead to more depleted vapour. Also, the idealised trajectory induces continuously saturated conditions, which is only to some extent captured by the diagnosed condensation onset temperature. By calculating the fractionation separately for each uptake event, we treat a potentially non-linear process as linear, which may only be valid to a first approximation. The full transport and fractionation history of an air mass could be considered by running MCIM as a box model along backward trajectories from reanalysis data (Helsen et al. 2005b). As another potential influence appears the assumption, that the initial isotopic composition of the moisture is taken from the lowest model level, while water vapour at the actual (higher) air parcel location may be already more depleted. This is in particular the case for the $\sim$26% of ABL moisture, which should clearly be more depleted than water vapour in the BL. If considered, this would lead to a further depletion of the modelled isotope values.

The fact that isotope values obtained in experiment 2 are considerably too high is obviously caused by the different water vapour initialisations. As the closure equation applied for modelling evaporation in the MCIM model is only valid on a global scale, it also fails to produce realistic initial values from the diagnosed RH, SST, and wind velocities. Jouzel and Koster (1996) noted from δD observations that the moisture source temperatures should be at least 5 K lower than assumed in the original MCIM model (15–20°C). We diagnose here average moisture source temperatures of 5.4–10.2°C, which is even more drastically lower than current assumptions. Hence, at the moment, there is probably no alternative to using GCM data as initial values for such Lagrangian modelling studies.

The fractionation conditions during transport and the water vapour's initial isotopic composition from the ECHAM4 simulation could be used to re-tune MCIM's internal microphysical parameters, and to make it fit better to the moisture sources and fractionation temperatures from this work. In comparison with observational data from Greenland ice cores and snow pits as in the preliminary study presented here, MCIM could then be used on a more physically constrained basis.

Finally, it is noteworthy that a recent Lagrangian approach to isotope modelling in Antarctica, both for single sites (Helsen et al. 2004, 2005a,b) and the whole of Antarctica (Helsen 2005) has provided simulations with the same model that showed good absolute correspondence with data. This good correspondence was however only achieved with several adjustments to the MCIM model. First, moisture origin was assumed to be identical with the air parcel location 5 days before arrival over Antarctica, and initialised with isotope ratios that were adjusted to yield good correspondence with observations (Helsen et al. 2004) or from a 3D mean isotope field from the ECHAM4 GCM (Helsen et al. 2005b). The calculation method applied by Helsen et al. (2005b) takes an intermediate position between isotope GCM and Lagrangian calculations: isotope ratios from the GCM mean field are incorporated in the calculation whenever moisture increases during a time step (termed *isotopic recharge*, see also Kavanaugh and Cuffey (2003)). It is probable that the well-tuned isotope GCM mean field thereby dominates the isotopic composition of the arriving back-trajectories. This issue may be further examined in the future (M. Helsen, *pers. comm.*, 2006).

4.6 Relative importance of fractionation temperatures

It is now attempted to draw a comprehensive picture of the relative importance of the diagnosed moisture sources and fractionation parameters for the isotopic results. The primary concern hereby is the origin of the NAO variability in the isotopic signal.

The mean changes of SST, T_{cnd}, T_{arr}, and T_{dif} with NAO phase are compared qualitatively[5] to identify the relative influence of the various parameters (Table 4.2). Tem-

[5]All absolute values are not significantly different, as they show large variability across the Greenland

Table 4.2: Comparison of the Greenland mean temperature parameters (top panel) and δ^{18}O influences (bottom panel). See text for details.

Variable	NAO+ (°C)	NAO− (°C)	Δ_T (K)
SST	5.4 ± 6.5	10.2 ± 6.8	+4.8
T_{cnd}	-1.4 ± 7.0	4.5 ± 6.5	+5.9
T_{arr}	-22.2 ± 3.9	-18.0 ± 4.4	+3.8
T_{dif}	20.7 ± 7.1	22.5 ± 7.1	−1.8

Variable	NAO+ (‰)	NAO− (‰)	Δ_δ (‰)
$\delta^{18}O_0$	-18.2 ± 2.9	-16.6 ± 2.8	+1.6
$\delta^{18}O_{EC}$	-23.2 ± 6.3	-19.8 ± 6.2	+3.4

perature differences Δ_T and isotope differences Δ_δ are used as indicators that are defined here so that positive (negative) values denote less (more) isotopic depletion during NAO+ than during NAO−.

Decreasing Greenland mean moisture source SSTs during NAO+ compared to NAO− months act towards more depleted isotope values (Table 4.2). Colder average T_{cnd} and T_{arr} during NAO+ months signify that fractionation begins and ends at lower temperatures (i.e. takes place at a lower temperature regime), which causes again more depletion. T_{dif} in contrast shows an inverse influence; for this parameter, the range of fractionation is smaller during NAO+ months, which leads to less depletion than for NAO− months. In a nutshell, there is more fractionation due to colder sources and a colder temperature regime during NAO+ than during NAO−months, while the smaller temperature difference, probably due to smaller transport distance and smaller temperature contrasts in the NAO+ phase, acts towards a reduced isotopic NAO signal.

The above interpretation of the temperature influences can now be utilised to understand the NAO variability of the isotopic composition of Greenland precipitation (Table 4.2). As the initial isotopic composition of the water vapour at the moisture source is known, a tentative quantification of the source and transport influences can be accomplished: SST differences of 4.8 K are associated with a change of 1.6‰ in the initial conditions of δ^{18}O. Considering the modelled average Δ_δ over Greenland of 3.4‰, this signifies that the combined transport effects contribute a net change of another 1.8‰. This leads to the conclusion that the NAO variability observed in δD and δ^{18}O in Greenland precipitation is a *combined* signal of moisture source and transport temperature changes, with roughly equal contributions.

The spatial interdependence of the fractionation regime (T_{cnd} and T_{arr}) and the fractionation range (T_{dif}) changes on the Greenland plateau for the three NAO mean conditions are investigated in Fig. 4.11. Mean values at all 284 surface points were used

plateau. A more regional interpretation could yield significant differences.

in this correlation, which is equal to examining all individual air parcels weighted by precipitation. A horizontal shift in this diagram signifies fractionation range changes; a vertical shift denotes fractionation regime changes. While the fractionation range is rather similar for the three phases, the fractionation regime is clearly colder during NAO+. The negative slope of the linear regression shows for all three phases that colder (higher) regions of Greenland are associated with a larger T_{dif}. A change in slope signifies that NAO variability has spatially non-uniform influences on the plateau. NAO+ months exhibit lower gradients ($-0.14\,\mathrm{K}\,\mathrm{K}^{-1}$) than NAO− ($-0.52\,\mathrm{K}\,\mathrm{K}^{-1}$). Consistent with Fig. 4.9 one would therefore expect a larger NAO variability at lower-altitude sites.

4.7 Further issues

Two further issues are briefly considered, namely the isotope-temperature relationship, and influences on the kinetic fractionation.

4.7.1 The isotope-temperature relationship

The isotope-temperature relationship is an empirical correlation between annual mean isotope ratios and surface temperatures on large spatial scales (see Chapter 2). This linear relationship is valid on multi-annual time scales. For climatic interpretations, it

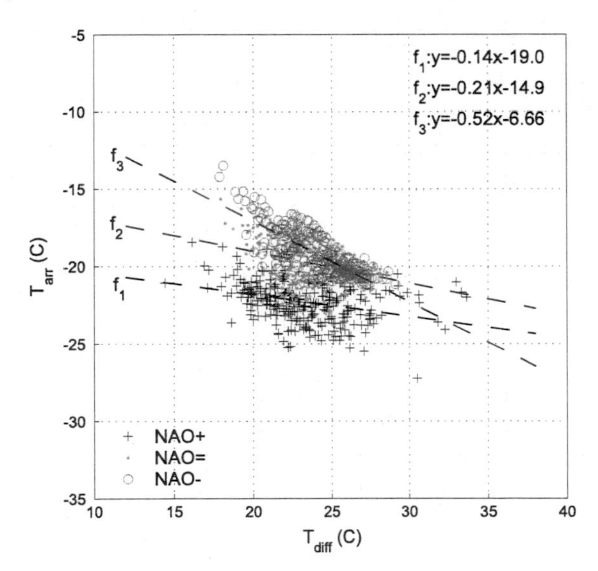

Figure 4.11: Fractionation regime shifts during the three NAO phases. The scatter plot shows the dependency of the phase mean arrival temperature on the difference temperature for all 284 arrival locations over Greenland. Dashed lines (f_1, f_2, f_3) are linear regressions.

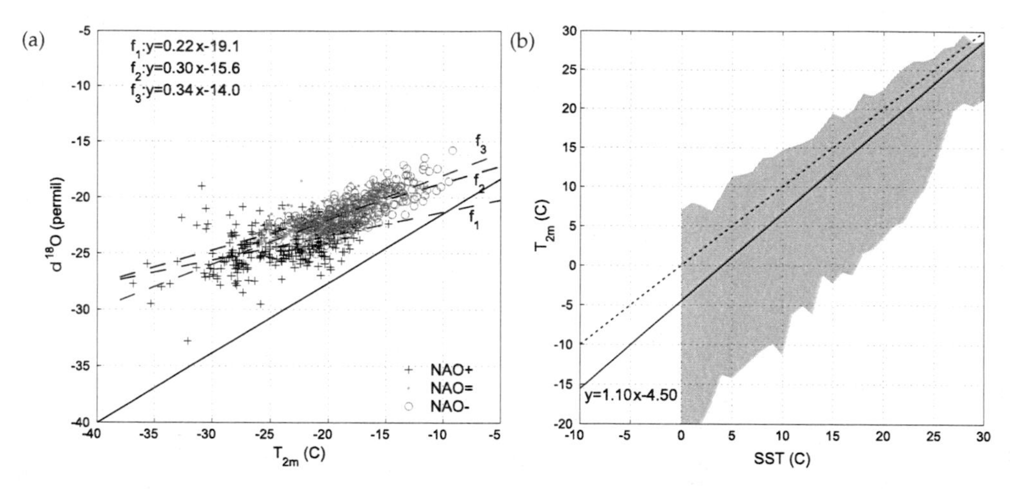

Figure 4.12: (a) Isotope-temperature dependency over Greenland for the three NAO phases. Dashed lines (f_1, f_2, f_3) are linear regressions. Solid line is Dansgaard's (1964) global isotope-temperature relationship (Eq. 2.14). (b) Scatter plot of T_{2m} vs. SST at the moisture uptake locations for the NAO− months. Shaded area gives the maximum and minimum range, solid line is a linear regression, the dashed line a 1:1 reference line.

requires a number of assumptions which allow for an extrapolation in time, for example a fixed relation between the arrival temperature and temperature difference, and a constant seasonality of precipitation. The data acquired from this study allow to asses some of the assumptions surrounding this empirical tool, and whether it is influenced by NAO variability.

Figure 4.12a shows the isotope-temperature relationship as a scatter plot of the modelled δ^{18}O versus T_{sfc} on the Greenland plateau during precipitation periods[6]. Data are phase mean values from all 284 surface grid points. Most data points fall on a range covered by Dansgaard (1964)'s empirical δ^{18}O-T relation (Eq. 2.14). Linear regressions for the three NAO phases show a decreasing offset and increasing slope with decreasing NAO phase. Generally, slopes are too low, which is probably due to using the warm-biased T_{sfc} (see Sec. 4.4.1), and the underestimated isotopic depletion. Conversely, a change in the ratio of precipitation to non-precipitation days could hence influence the isotope-temperature relationship. Note also that this study is limited to conclusions on the winter season, in contrast to the annual mean considerations, e.g. in Dansgaard (1964).

In future studies with a better tuned MCIM model, further questions could be addressed, such as whether the isotope-temperature relationship holds on short (synoptic) time scales, and if and how it varies with the NAO for other seasons.

[6]A warm bias is introduced here in comparison to winter mean temperatures.

4.7.2 Transport influences on kinetic fractionation

As a final issue, the temperature contrast between air and water at the evaporation sites is investigated. As was found in Sec. 3.5.4, most air parcels resided over the continents prior to the first moisture uptake. Figure 4.12b demonstrates that the air-sea temperature contrast increases with decreasing SST, which is consistent with a stronger land-sea temperature contrast in northern latitudes during winter. It can be speculated that this non-uniform temperature contrast could influence non-equilibrium fractionation processes during evaporation. Cappa et al. (2003) found that surface cooling in the uppermost 0.5 mm of a water surface due to evaporation influences the fractionation of D and ^{18}O differently, a fact which is currently not included in the evaporation parametrisation of GCM isotope models. The larger temperature contrast found here could lead to the maintenance of a larger surface cooling during evaporation, and hence cause currently neglected latitudinal differences in d-excess in the water vapour near the surface. This issue could be investigated further during the development and evaluation of evaporation parametrisations based on the findings of Cappa et al. (2003), and further observational data from the marine BL.

4.8 Concluding remarks

The Lagrangian isotope modelling results presented in this chapter have potential implications beyond the focus of the present work:

- In an initial comparison with observational data, the Lagrangian approach to model the isotopic fractionation of winter precipitation in Greenland from stable isotopes did not produce a realistic level of isotopic depletion. However, it captured the NAO variability from the diagnosed input parameters to a surprisingly high degree. Based on these results, the stable isotope variability of δD and $\delta^{18}O$ can be understood as a combined effect of both source region changes and changes in transport and arrival conditions, which are both induced by the NAO. Local SST changes in comparison seem to have negligible influence. Results were much less conclusive for the d-excess. Here, the isotopic values of d-excess decreased during fractionation. A sensitivity experiment confirmed the findings of Jouzel and Koster (1996) that GCM output should be used for the initialisation of MCIM with δD and $\delta^{18}O$.

- Taking the NAO as an analogue situation for longer-term climatic shifts (c.f. Chapter 2), it can be concluded from this work that stable isotope ratios are not plain site temperature signals, but partially reflect dynamically induced changes in the general circulation, which includes the atmospheric water cycle. The role of moisture sources for the interpretation of ice cores has recently been recognised by Masson-Delmotte et al. (2005a,b). Rapid shifts in isotopic composition observed in paleo-records could in addition to temperature also partly reflect such circulation changes.

- Spatial gradients of variability were found over the Greenland plateau. Since the spatial resolution of the present isotope model results is available at a reasonably high degree and based on reanalysis data, it is now possible to select and compare sites where the largest NAO signal can be expected, such as the very north, or the southern half of the plateau. This could support studies which aim to reconstruct NAO variability from isotopic signals (Vinther et al. 2003) or accumulation series (Mosley-Thompson et al. 2005).

- Seasonality of precipitation is an important component of inter-annual isotope variability. As this study was only concerned with winter months (which is when the NAO has its strongest imprint), the influence of seasonality changes, both in space and time, is completely neglected here. Nevertheless, seasonality changes in the precipitation regime can have major impacts on the annual mean isotope signal, and should be an important component of future research efforts.

Further refinement of the approach presented here, in particular with respect to adjusting the MCIM model to the newly identified fractionation conditions, and an extension to a full seasonal cycle, are expected to lead to fruitful future applications of this Lagrangian isotope modelling approach.

病起見閑雲

病起見閑雲空中聚又分滯留堪笑我舒卷不如君觸
石終無跡從風或有聞仙山足鸞鳳歸去自同群

Recovering, I watch the idle clouds
crowding together in the sky, then parting,
pausing long enough to laugh at me–
I could never match your twists and turnings!
Colliding with rocks, you leave no trace;
obeying the wind, you seem to have ears.
This mountain of the immortals has phoenixes enough–
better be off to flock with your own kind!

Ch'i-chi (864–937), translated by Burton Watson
in "The clouds should know me by now"

Part II

Eulerian approach

Chapter 5

A Water Vapour Tagging (WVT) Methodology

In this chapter, tagging approaches for tracing the movement of a marked amount of water vapour (WV) through model space and time are described. Then, based on a number of 1D experiments, an implementation strategy for the 3D regional climate model CHRM (Climate High Resolution Model, see Chapter 6 for a detailed description) is chosen. While natural tracers in geophysical systems are not always readily accessible, in the environment of a numerical model, arbitrary quantities originating from or passing through a specified area can be marked and then traced during their further movement in the geophysical model system. Such numerical mark-and-trace approaches are often referred to as *tagging* methods. This implies the association of a unique tag or label which is attached to the traced quantity. Common tagging approaches in NWP models typically involve the introduction of a subset of the quantity under consideration (here WV), which is advected independently from and if possible consistently with the main field as a (passive or reactive) tracer through the model domain.

Choosing the right advection numerics is an important ingredient for the modelling of tracer advection. As noted by Joussaume et al. (1986), positive definite advection numerics should be used for tracer advection to avoid problems with negative values, hence they used a forward scheme. Koster et al. (1986) used the slope scheme of Russell and Lerner (1981). Later, Koster et al. (1992) noted problems in the tracer field due to numerical noise produced by that advection scheme. Numaguti (1999) used a spectral method, but had to deal with negative values after the Fourier decomposition. Bosilovich and Schubert (2002) use a positive definite and conservative semi-Lagrangian method (Lin and Rood 1996), which however requires a remapping after each integration.

A second important requirement is to use consistent numerics for the tracer and the main prognostic moisture field, and to thereby keep differences due to numerical effects between the tagged quantity and the original field at a minimum. A seemingly straightforward way to achieve numerical consistency is to use the same numerical scheme for the tracer advection as for the advection of the total field. Another typical requirement for tracer applications is the conservation of mass. Spurious sinks and sources of the advected quantity due to numerical deficiencies should be excluded. Sometimes however, as is the case here, the requirement of conservation of mass may force one to deviate from the requirement of numerical consistency.

The numerics of the CHRM model treat water vapour in units of specific humidity q (kg/kg). However, as will be shown below, for a deformative flow field, specific humidity advection does not conserve water mass. The conservation of WV mass requires a formulation in units of water vapour density ρ (kg/m^3) or water mass m_w (kg) instead. Conservative advection of these two quantities requires a different discretisation than the advection of a specific humidity. Hence a mass-conservative water vapour tagging method for the CHRM model will be numerically different from the CHRM model numerics which advects q. In order to assess the errors introduced thereby, idealised numerical simulations are carried out for different tagging methods.

5.1 Tagging numerics

In the following subsections, first some basic and advanced numerical concepts will briefly be reviewed. On this basis, three different tagging methodologies will be described. First, a fully numerically consistent but non-conservative approach will be presented, followed by a tagging algorithm involving a conservative parallel tracer field. Finally, a new hybrid approach being nearly conservative and leading to more consistency between the main and tracer fields is proposed. The 1D model setup used for evaluating the performance of the three tagging methods will be given in Section 5.2. Following the results from a suite of numerical experiments presented in Section 5.3, a rationale for the chosen tagging implementation will be given in the final section of this chapter.

5.1.1 Basic numerics

Consider the one-dimensional advection equation for a quantity in the absence of sources and sinks:

$$\frac{Dq}{Dt} = \frac{\partial q}{\partial t} + \vec{u}\nabla(q) = 0 \,. \tag{5.1}$$

In one dimension, this reduces to

$$\frac{\partial q}{\partial t} + u\frac{\partial q}{\partial x} = 0 \,. \tag{5.2}$$

In the case of a constant $u > 0$, this equation can be discretised in space and time on a regular grid using the finite increments Δx and Δt (e.g. Schär 2002):

$$\frac{q_i^{n+1} - q_i^n}{\Delta t} + u\frac{q_{i+1}^n - q_i^n}{\Delta x} = 0\,. \tag{5.3}$$

This is the so-called upstream forward scheme. When rearranged, Eq. (5.3) provides a calculation rule to derive q^{n+1} from q^n:

$$q_i^{n+1} = q_i^n + \alpha(q_{i+1}^n - q_i^n) = 0\,. \tag{5.4}$$

Here, $\alpha = u\frac{\Delta t}{\Delta x}$ is the *Courant* number, an important parameter considering the stability of a numeric scheme. Due to its highly diffusive properties, the upstream scheme is of limited practical use. In the CHRM model, the advection equation (Eq. 5.1) is discretised by means of the time centered leap-frog (LF) scheme:

$$q_i^{n+1} = \overline{q_i^{n-1}} + 2\alpha(q_{i+1}^n - q_{i-1}^n)\,. \tag{5.5}$$

The amplitude error of the leap-frog discretisation is considerably smaller than that of the upstream forward scheme. However, the leap-from scheme is associated with a large phase error, which leads to the advection of short-wave anomalies in opposite to the direction of the advection velocity (Schär 2002). In addition, leap-frog numerics are associated with non-physical oscillations due to the so-called numerical mode. The numerical mode is usually damped by means of an *Asselin* filter, denoted by the overbar in Eq. (5.5):

$$\overline{q_i^n} = q_i^n + \gamma(\overline{q_i^{n-1}} - 2q_i^n + q_i^{n+1})\,. \tag{5.6}$$

This temporal filter operation is performed at the end of each time step. The filter parameter γ is typically chosen in the range of 0.06 to 0.25; the CHRM numerics use $\gamma = 0.15$.

In the CHRM model, as in most NWP models, the atmosphere is discretised on a staggered grid. On staggered grids mass variables (q, m, T) are defined at the centre of a cell, while flux variables (u, mass fluxes F) are defined at the cell borders (Fig. 5.1). Positions at cell borders are denoted by indices of one half (e.g. $q_{n+1/2}$). On a staggered grid with a non-uniform wind field, the specific humidity advection equation in leap-frog discretisation is

$$q_i^{n+1} = \overline{q_i^{n-1}} + \frac{2\Delta t}{\Delta x}\left(\frac{1}{2}u_{i+1/2}^n(q_{i+1}^n - q_i^n) + \frac{1}{2}u_{i-1/2}^n(q_i^n - q_{i-1}^n)\right). \tag{5.7}$$

For a uniform flow field, Eq. (5.7) reduces to Eq. (5.5). This discretisation in its 2D counterpart is used in the CHRM model for the advection of the WV specific humidity field q.

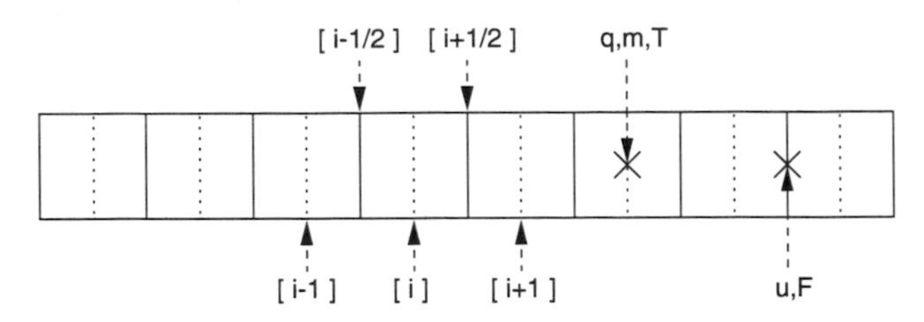

Figure 5.1: Setup of the staggered grid in the CHRM model in 1D. Solid lines denote the staggered variable positions (for u, F), while dotted lines indicate the mass center positions of the cells (for q, m, T).

A further advanced numerical discretisation is the Lax-Wendroff (LW) scheme. Here, the advected quantity q^n is determined from quadratic interpolation on the spatial interval $[x_{i-1}, x_{i+1}]$ by

$$q^n_{i+\mu} = a\mu^2 + b\mu + c \tag{5.8}$$

at a position μ (Schär 2002). The parameters a,b, and c are given by

$$\begin{aligned} a &= \frac{1}{2}\left(q^n_{i+1} - 2q^n_i + q^n_{i-1}\right) , \\ b &= \frac{1}{2}\left(q^n_{i+1} - q^n_{i-1}\right) , \\ c &= q^n_i . \end{aligned}$$

The Lax-Wendroff discretisation of Eq. (5.2) for a constant $u > 0$ is then

$$q^{n+1}_i = q^n_i - \frac{\alpha}{2}\left(q^n_{i+1} - q^n_{i-1}\right) + \frac{\alpha^2}{2}\left(q^n_{i+1} - 2q^n_i + q^n_{i-1}\right) . \tag{5.9}$$

5.1.2 Conservation of mass

Conservation of mass is a prerequisite for many tracer applications, but by no means inherent to all discretisations. For WV, mass conservation can be expressed by means of the continuity equation:

$$\frac{\partial \rho}{\partial t} + \nabla \cdot (\vec{u}\rho) = 0 . \tag{5.10}$$

Eq. (5.10) cannot be formulated in units of specific humidity (given a non-uniform wind field). One practical solution is to convert a specific humidity field to WV density or mass before a mass conservative advection scheme is applied:

$$\rho = \frac{q \cdot p}{R \cdot T} \tag{5.11}$$

$$\text{or } m = \frac{q \cdot m_a}{q} = \frac{q \cdot \Delta x \Delta y \Delta p}{q \cdot g} \tag{5.12}$$

where m_a is the mass of a volume of moist air, and g is the acceleration due to gravity. In pressure coordinates, the height of a grid cell may change according to the flow field at each time step, hence the water masses have to be calculated anew for each time level and each grid cell. In the following, the discussion will be limited to formulations using the water vapour density. The findings can however without restrictions be extended to water mass as well.

A rational way to formulate mass conservative numerical schemes is to discretise Eq. (5.10), leading to so-called *flux-form* schemes. The flux-form discretisation of the leap frog scheme reads

$$\rho_i^{n+1} = \rho_i^{n-1} - \frac{2\Delta t}{\Delta x}\left(F_{i+1/2}^n - F_{i-1/2}^n\right), \tag{5.13}$$

$$\text{where } F_{i-1/2}^n = u_{i-1/2}^n \cdot \frac{1}{2}\left(\rho_{i-1}^n + \rho_i^n\right) \tag{5.14}$$

$$\text{and } F_{i+1/2}^n = u_{i+1/2}^n \cdot \frac{1}{2}\left(\rho_i^n + \rho_{i+1}^n\right) \tag{5.15}$$

approximate the fluxes at the cell borders (Fig. 5.1). Most common numerical schemes can be formulated in flux-form, such as the forward upstream, leap frog, and Lax-Wendroff discretisations (Schär 2002).

Flux form discretisations are designed to preserve total mass, but to some extent this is achieved by physically questionable means. The LF and LW scheme can only compensate the mass gain in overshoots by undershoots, which represent physically meaningless negative mass (see Section 5.1.5). Negative values during advection can be avoided by offsetting the advected field uniformly. The difficulties however reappear when the offset has to be subtracted from the advected field, for example when values serve as input for physical parameterisations.

5.1.3 Positive definite non-oscillatory schemes

Another way to avoid the occurrence of negative values for non-negative quantities like mass is to use numerical schemes which do not produce negative values, or in more general terms, preserve the sign of the advected quantity. Preservation of the sign is hence an important requirement for the physical consistency of numerical schemes. So-called *positive definite* advection schemes achieve preservation of sign by means of flux correction techniques (see below). In addition, flux correction can be used to avoid both overshoots and undershoots, and to derive monotonicity-preserving schemes (so-called *non-oscillatory* schemes).

A widely used numerical advection algorithm with positive definite and non-oscillatory properties is the MPDATA scheme (multidimensional positive definite advection transport algorithm with non-oscillatory option) by Smolarkiewicz and Grabowski (1990) (see also Smolarkiewicz and Margolin (1998) and Smolarkiewicz (2005) for further details). It combines the small amplitude error of the of the Lax-Wendroff scheme with the small phase error of the upstream forward scheme. The advection calculation is performed iteratively. In a first step, a donor cell is determined by a simple upstream differencing, which is only first-order accurate. Then, in a second and optional third iteration, this first-order error is reduced by means of a Lax-Wendroff type discretisation, leading to an increasingly accurate solution of the advection equation. Technically, a flux correction is implemented in the MPDATA scheme. The corrective measures are achieved by using modified wind velocities ('pseudo-velocities'), which do not have physical significance. Offering both preservation of sign and a non-oscillatory option, the MPDATA algorithm is the most advanced numerical schemes considered here.

5.1.4 Explicit numerical diffusion

The advection equation (Eq. 5.1) is often extended by a diffusive term, which represents diffusive effects on small scales that occur during transport:

$$\frac{Dq}{Dt} = \frac{\partial q}{\partial t} + \vec{u}\,\nabla q - \kappa_4 \nabla^4 q = 0\,. \tag{5.16}$$

Here, κ_4 is the 4^{th}order diffusion coefficient, with a value depending on the spatial and temporal discretisation intervals. For stability reasons, the numerical diffusion coefficient is limited as

$$\kappa_4 = \frac{\Delta x^4}{2\pi^4 \Delta t}\,. \tag{5.17}$$

In numerical modelling, the diffusion term in Eq. (5.16) has two purposes: First, it represents the physical process of small scale mixing, and second, it is used as a numerical filter to damp short-wave numerical noise. This second purpose is of particular importance when the leap-frog scheme is applied, as the numerical diffusion is strongly scale-selective: anomalies with wavelength of $2\Delta x$ are reduced by 66%, while the damping for $10\Delta x$ is only 0.6%. Thus, numerical noise due to the high dispersion of the leap frog discretisation can be effectively reduced. In a one-dimensional case, the equation for the 4th-order diffusion is

$$\frac{\partial q}{\partial t} - \kappa_4 \frac{\partial^4 q}{\partial x^4} = \frac{\partial q}{\partial t} - \kappa_4 \frac{\partial^2 q}{\partial x^2}\left(\frac{\partial^2 q}{\partial x^2}\right) = 0\,. \tag{5.18}$$

In discretised form, and rearranged for calculating purposes, Eq. (5.18) becomes

$$q_i^{n+1} = q_i^n - \frac{\Delta t}{\Delta x^4} \cdot \kappa_4 (q_{i+2}^n - 4q_{i+1}^n + 6q_i^n - 4q_{i-1}^n + q_{i-2}^n) \, . \tag{5.19}$$

From Eq. (5.19), it becomes obvious that the filter involves a neighbourhood of ± 2 grid boxes. Therefore, near the domain boundaries in the CHRM model, a 2nd-order numerical diffusion is applied instead.

5.1.5 Advection tests of the numerical schemes

The advection and diffusion schemes presented above are compared with two standard advection tests, and their characteristic properties are presented. First, a 9 cells wide rectangular initialisation (Type I) is advected with the leap-frog, Lax-Wendroff, and MPDATA numerical schemes, both including and excluding a 4th-order numerical diffusion. Then, the same algorithms are compared for a 9 cells wide zigzag initialisation (Type II) of the same width.

In the Type I advection tests, it is apparent that the LF numerics produce high-frequency anomalies travelling upstream, even though damped by an Asselin filter (Figs. 5.2a, b, thin line). This is due to negative group velocities occurring for short-wave anomalies in the LF scheme (Schär 2002, his Fig. 3.4a). Due to its scale-selective properties, the numerical diffusion effectively removes the small-scale anomalies, while keeping the large-scale features almost unchanged (Figs. 5.2a, b, thick line).

The LW scheme does not produce as much numerical noise as the LF scheme, but shows typical overshoots (values exceeding the initial value of a grid cell) and undershoots (here values with negative WV densities) upstream of large gradients (Figs. 5.2c, arrows). The numerical diffusion only slightly influences the advected field, with a small reduction of the strongest extremes. These drawbacks of the LF an LW schemes are in contrast with the small amplitude and phase errors, and their computational efficiency.

In contrast, the advection test using MPDATA numerics preserves the shape of the initialisation very accurately, and produces neither overshoots nor undershoots (Figs. 5.2e, f, thin line). Both, phase and amplitude errors are very small. However, when explicit numerical diffusion is implemented, the MPDATA solution actually becomes worse: overshoots are created at the maximum of the anomaly, which counteract the monotonicity-preserving properties of the MPDATA scheme alone (Figs. 5.2e, f, thick line).

Performing the advection tests with the Type II initialisation consolidates the discussion above. (Fig. 5.3). Now, due to the short-wave characteristics of the initialisation, the LF numerics produce a short-wave anomaly of the same amplitude as the residual long-wave anomaly, but with reversed group velocity (Fig. 5.3a, b). The numerical diffusion is able to damp this non-physical effect to some degree, but some numerical

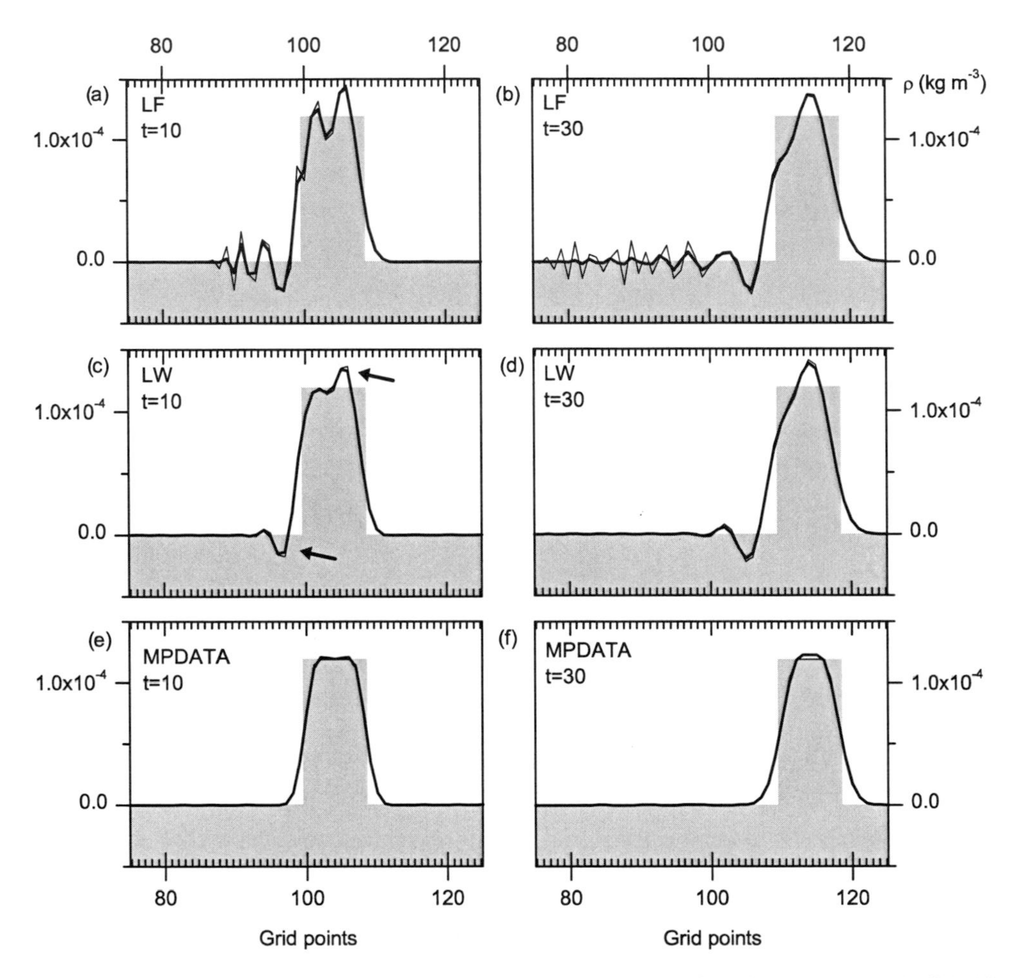

Figure 5.2: Advection of a rectangular initialisation with the leap frog (LF), Lax-Wendroff (LW), and MPDATA algorithms after 10 and 30 time steps, respectively. Shaded area gives the analytic solution, while the lines denote numerical solutions with (thin line) and without (thick line) numerical diffusion.

noise remains (Fig. 5.3b). In comparison, the LW scheme quickly removes the zigzag shape of the initialisation and produces again the characteristic over- and undershoots upstream of the strongest gradients (Fig. 5.3c, d). Again, the MPDATA solution is devoid of numerical noise. However, the diffusion of the MPDATA scheme rapidly averages out the initial zigzag anomaly (Figs. 5.3e, f). The explicit numerical diffusion has hardly any distinguishable influence on the LW and MPDATA advection tests with a Type II initialisation.

In summary, comparing the advection results for the LF and LW schemes with the MPDATA scheme exhibits some pronounced differences. Numerical noise due to group

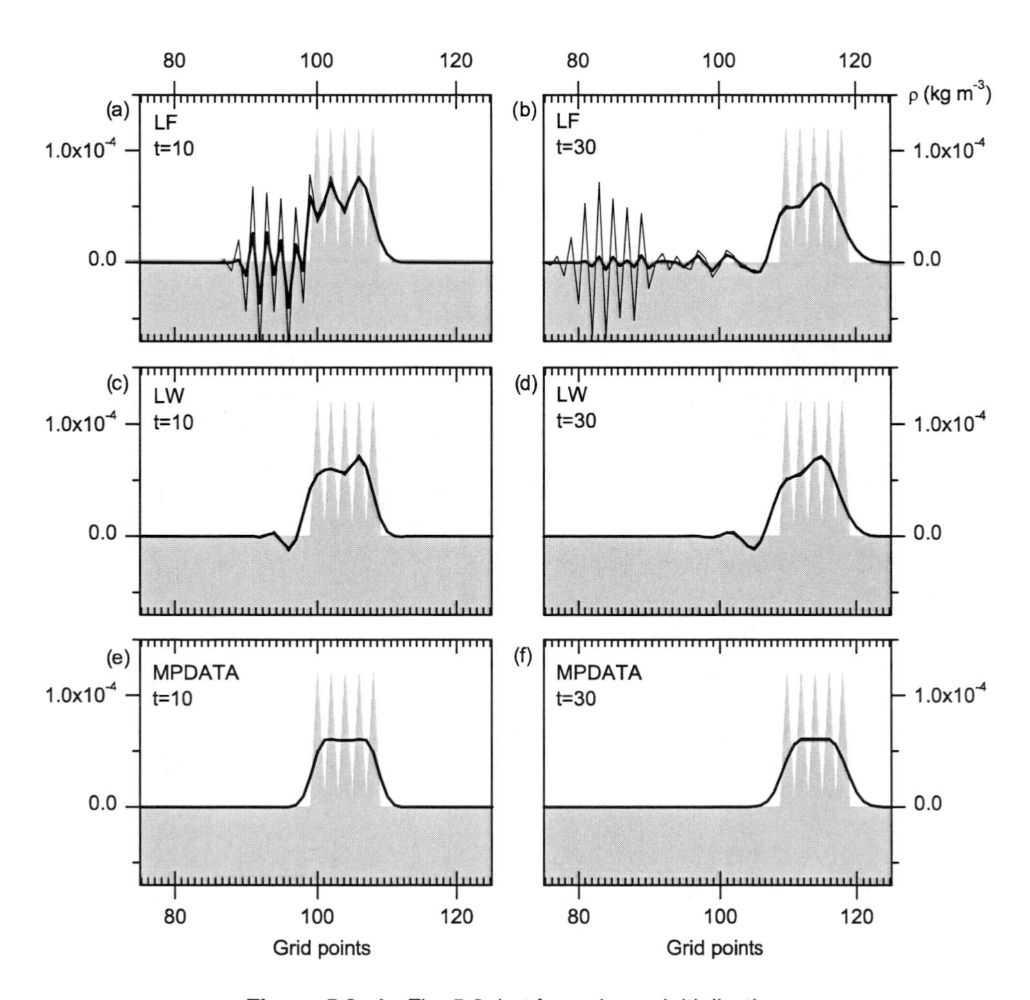

Figure 5.3: As Fig. 5.2, but for a zigzag initialisation.

velocity reversal or overshoots/undershoots could cause difficulties when the numerical schemes are applied in a tagging algorithm. In light of choosing numerics for a tagging algorithm, this means that choosing the same numerics for tracer and main field advection does not guarantee numeric consistency. In addition, the tagged quantities are characterised by sharper gradients and hence smaller-scale fluctuations than the main field. Hence, it is advisable to choose a more advanced numerical scheme with properties favourable for tagging applications, such as the MPDATA scheme. Explicit numerical diffusion improves the advection results for the LF and to a limited extent for the LW schemes, while it worsens the results for the MPDATA advection scheme. Explicit numerical diffusion should hence not be used with the advanced MPDATA scheme.

5.1.6 Tagging algorithms

The change in content of marked water $\Delta\rho_t$ (where the subscript t stands for *tagged*) from one time step to another is at each grid cell given by

$$\Delta\rho_t = \rho_t^{n+1} - \rho_t^n = \Delta\rho_a + \Delta\rho_d + \Delta\rho_p \tag{5.20}$$

where $\Delta\rho_a$, $\Delta\rho_d$, and $\Delta\rho_p$ are the changes in marked moisture due to advection, diffusion, and parameterised physical processes, respectively. While these three components can be addressed independently, the full tagging algorithm in the end needs to accommodate all three contributions. The discussion below will focus on the advection and diffusion components, $\Delta\rho_a$ and $\Delta\rho_d$. The physical parameterisations component $\Delta\rho_p$ poses particular problems. For example, thresholds for the triggering of convection may not work for a tagged subset of the main WV field. The handling of this third component $\Delta\rho_p$ is discussed in Chapter 6.

Three basic concepts of tagging algorithms are presented here. The first concept uses the same non-conservative numerics as the CHRM advection and diffusion scheme. After initialisation, the tagged WV is advected independently from the main field, hence it is termed here a *parallel* tracer. All previously published tagging implementations (Joussaume et al. 1986; Koster et al. 1986; Numaguti 1999; Bosilovich and Schubert 2002) in GCMs fall into this category. The second concept uses an advanced conservative numerical scheme, again in a parallel tracer implementation. Third, a new tagging concept is proposed, which takes an intermediate position between the first two approaches. Rather than being completely independent, at each time step the tagged field is synchronised with the main field. Hence, the tracer is termed here a *synchronised* tracer.

I. Parallel tracer with specific humidity numerics (PTQ)

The PTQ tagging approach uses the same numerics and units as are used for the advection of the main WV field. The tracer field q_t is initialised as a subset of the main WV field q. After initialisation, the field q_t has the same value as q in places where tagged WV exists, and equals zero at all other grid points. Using the same numerics as for q, q_t is then advected within the model domain. Optionally, numerical diffusion (Eq. 5.19) and an Asselin filter (Eq. 5.6) can be applied to the tagged WV field. Even though the PTQ tagging concept seems numerically consistent with the main WV field, stronger gradients at the boundary of tagged WV can still create numerical differences. In addition, the CHRM advection scheme for q does not conserve mass for non-uniform wind fields. For the full WV field this may be less of a limitation than for practical tagging applications, where conservation of mass matters.

II. Parallel tracer with water mass conservative numerics (PTW)

In the PTW tagging approach, the tagged WV field ρ_t is initialised by converting a subset q_t of the main WV field q from specific humidity to a tagged WV density field ρ_t using Eqs. (5.11 and 5.12). The tracer field ρ_t is then advected using mass flux numerics (Eqs. 5.13–5.15), either with the same or preferably a better discretisation scheme than for the main WV field. This approach has the important advantage of being mass conservative. As a drawback, the parallel tracer field, after a number of iterations, has evolved along a different trajectory than the main WV field q, in particular when a different numerical scheme is applied.

Hence, the PTQ and PTW tagging algorithms involve the following steps:

1. Initialise the tagged moisture field q_t^n, ρ_t^n
2. Advect the tagged WV fields q_t^n, ρ_t^n using Eqs. (5.6 and 5.7) or Eqs. (5.13–5.15) to derive q_t^{n+1}, ρ_t^{n+1}
3. Optional: Calculate the tagged water fluxes due to numerical diffusion
4. Collect the tagged water fluxes due to physical parameterisations (see Chapter 6)
5. Advance to next time step

III. Synchronised tracer with water mass conservative numerics (STW)

The STW tracer proposed here aims at incorporating the advantages of both the PTQ and the PTW approach. Instead of introducing a new field for a tracer quantity, only the *fraction* of tagged moisture $f \in [0, 1]$ relative to the total moisture content is stored for each grid cell:

$$f = \frac{\rho_t}{\rho}. \tag{5.21}$$

Given the field of these fractions f, the tagged moisture field ρ_t can easily be derived from ρ at each time step. In addition, the fraction of marked water for each cell is an information which is required anyway for handling the contributions from the parameterisations (Eqs. 5.20 and $\Delta\rho_p$) to the total tagged WV transport. While not being restricted to any particular numerical scheme, the STW advection equation is given here for a 1D leap-frog discretisation:

$$\rho_i^{n+1} \cdot f_i^{n+1} = \rho_i^{n-1} \cdot f_i^{n-1} - \frac{2\Delta t}{\Delta x}\left(F_{i+1/2}^n - F_{i-1/2}^n\right), \tag{5.22}$$

$$\text{where } F_{i-1/2}^n = u_{i-1/2}^n \cdot \frac{1}{2}\left(\rho_{i-1}^n \cdot f_{i-1}^n + \rho_i^n \cdot f_i^n\right) \tag{5.23}$$

$$\text{and } F_{i+1/2}^n = u_{i+1/2}^n \cdot \frac{1}{2}\left(\rho_i^n \cdot f_i^n + \rho_{i+1}^n \cdot f_{i+1}^n\right) \tag{5.24}$$

are the fluxes from left and right into a cell. Up to this point, the STW numerics are mathematically equivalent to the PTW approach, only that ρ_r has been replaced by $\rho \cdot f$ (compare Eqs. 5.13–5.15). Hence, rearranging Eq. (5.22) gives the term $\Delta\rho_a$ in Eq. (5.20). Optionally, numerical diffusion can then be calculated for the tagged WV field using Eq. (5.19).

As a final step, the new tagged WV density field ρ_t^{n+1} is reconverted into a fraction of the total WV density field ρ^{n+1}. This total field is again calculated from converting the total WV specific humidity field q^{n+1} according to Eqs. (5.11 and 5.12). At this point, inconsistencies between the tagged an the main WV field can become evident, which can either be due to different discretisations, numerical noise, or changes from physical parameterisations to the main field. In such a case, the fraction f could become greater than 1.0 or smaller than 0.0 (Eq. 5.21). Then, corrective measures have to be taken, namely imposing upper ($f \equiv 1$) and lower ($f \equiv 0$) limits. Obviously, this would result in the destruction or production of tagged water mass, and mass would not be conserved. On the other hand, an important advantage of the STW approach becomes evident: The reconversion of the tagged moisture field into fractions ensures that the amount of tagged water never exceeds the total amount of water in a grid cell. Thus, some degree of synchronisation between the tracer field and the main water vapour field is ensured, however at the cost of potentially incomplete conservation of mass. The examples in Section 5.3 will illustrate these aspects of the STW approach.

In summary, the proposed tagging algorithm STW involves the following steps:

1. Calculate q^{n+1} from the usual advection algorithm (here a leap frog scheme)
2. Transform the specific humidity fields $q^{n-1,n,n+1}$ for all time levels into water vapour density fields $\rho^{n-1,n,n+1}$
3. Reconstruct the tagged water vapour fields $\rho_t^{n-1,n}$ from $\rho^{n-1,n}$ using Eq. (5.21)
4. Advect the tagged water vapour field ρ_t^n to derive ρ_t^{n+1} using Eq. (5.22)
5. Transform the tagged water vapour fields ρ_t^{n+1} back into fractions f^{n+1} using Eq. (5.21)
6. Optional: Calculate the water fluxes due to numerical diffusion

7. Collect the water fluxes due to physics parameterisations (see Chapter 6)

8. Advance to next time step

5.2 Idealised 1D advection experiments

In order to evaluate the performance and applicability of the tagging algorithms presented above, a series of idealised advection experiments has been carried out in a one-dimensional model environment. The model setup, the applied evaluation criteria, and the results are presented in the following sections.

5.2.1 Model environment and setup

The model domain is set up as a 1D staggered grid with cyclic boundaries (Fig. 5.1). Specific humidity, water vapour density, pressure, and air temperature are defined at the cells' mass points, while fluxes between cells and wind velocities are located at the staggered grid points. Each cell represents a volume of air with the dimensions given in Table 5.1. For a first set of experiments, the model domain was initialised with a uniform wind field, a second series of tests involved a sinusodially distorted wind field (Fig. 5.4).

Three different initialisations of total and tagged water vapour were used for the experiments (Fig. 5.5). Tagging a contiguous section of a homogeneous main WV field resulted in a rectangle-shaped initialisation of the tracer field (Fig. 5.5a, Type I). Tagging the water vapour in every second cell within a section of a homogeneous main WV field results in a pulsed initialisation of the tracer (Fig.5.5b, Type II). In a real-world model study, such a pulsed initialisation could for instance result from diurnal cycles in evaporation at a tagged source region (even though such cycles would typically act

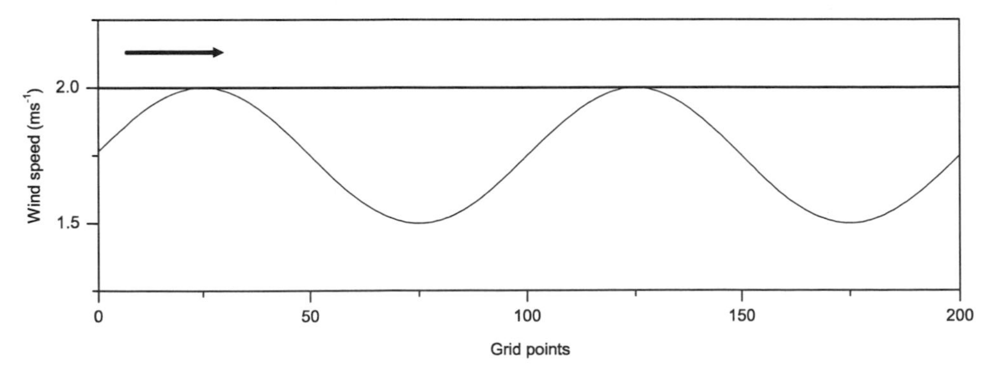

Figure 5.4: Wind field for the uniform (thin line) and oscillating (thick line) wind field experiments. The arrow indicates the direction of positive wind velocities.

Table 5.1: Parameters of the 1D model setup.

Parameter	Value	Parameter	Value
No. of cells	200	Air temperature	283 K
Time step	50 s	Wind speed (fixed)	$2\,\mathrm{m\,s^{-1}}$
Cell width	200 m	Wind speed (variable)	$1.5\text{–}2.0\,\mathrm{m\,s^{-1}}$
Cell height	50 hPa	Specific humidity	$0.1\,\mathrm{g\,kg^{-1}}$
Pressure	1000 hPa		

at larger time intervals). In a third setup, the main field itself has a pulsed structure (Fig. 5.5c, Type III). This last setup serves to test the tagging mechanisms in situations where numerical effects cause distortions in the main field. In particular, the PTW and STW methods could give different results in this setup. In all setups, the initialisation section extends over 9 grid cells out of the total 200 grid cells.

The PTQ tracer field was initialised with the same specific humidity as the main WV field in cells where tagged WV was present, and zero specific humidity at all other locations. The PTW tracer field was initialised in the same way, except that the specific humidity was converted to WV density using Eq. (5.11). Finally, the STW tracer field was initialised with $f = 1.0$ where tracer was present, and $f = 0$ in all other cells. This leads effectively to identical initialisations for the PTW and STW tracers at time $t = 0$. Note that in this 1D study only a single water vapour tracer was implemented. As the WV tracers are passive, implementing more tracers would however not lead to different results.

The experiments were carried out with different numerical advection algorithms. As in the CHRM model, the main WV field q and the tracer PTQ were always advected using leap-frog numerics according to Eqs. (5.5 and 5.6) with an Asselin filter coefficient $\gamma = 0.15$. The tagging methods PTW and STW were tested using a range of numerical schemes, including the upstream-forward, leap-frog, Lax-Wendroff, and MPDATA

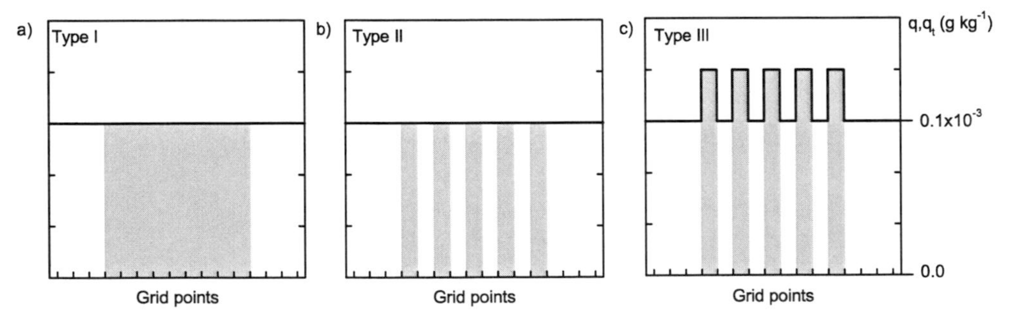

Figure 5.5: Different tagging test initialisations. The solid line gives the water vapour density of the main field, the shaded areas indicate the tagged water vapour. (a) Rectangle tagged area in homogeneous main field (Type I), (b) Pulsed initialisation in homogeneous main field (Type II), (c) Pulsed initialisation with a pulsed main field (Type III).

schemes. Here, only the results for the latter three will be presented. Mass-conservative flux forms were used in the PTW and STW advection numerics.

5.2.2 Evaluation criteria

A number of evaluation criteria were defined in order to compare the performance of the tagging methodologies. A common approach to compare numerical schemes is to compare against an analytical solution. Here, the tracer fields were compared by examining the differences between the numerical and analytical solutions at a given instant of time. Takacs (1985) introduced a more quantitative measure of algorithm performance. He identified three useful quantities to measure the errors of a numerical scheme: the total error E_{tot}, which is composed of the dissipation error (E_{diss}) and dispersion error (E_{disp}) according to

$$E_{tot} = E_{diss} + E_{disp}. \tag{5.25}$$

The dissipation error E_{diss} quantifies the energy dissipation for different wavelengths. The LF scheme for example has a large dissipation error, which results in anomalies at different wavelengths travelling at different speeds, partly even in opposite directions. The dispersion error E_{disp} represents the diffusive spread of the advected quantity. The upstream scheme, for example, has a zero E_{diss} and a large E_{disp}, resulting in a uniform spread of the advected anomaly.

The three quantities E_{tot}, E_{diss}, and E_{disp} are calculated here to quantify the performance of the two methods. According to this approach, the total error E_{tot} is defined as the mean square error (Takacs 1985):

$$\begin{aligned} E_{tot} &= \frac{1}{M}\sum_{j=1}^{M}(q_{T_j} - q_{N_j})^2 && (5.26)\\ &= \sigma^2(q_T) + \sigma^2(q_N) - 2\,\mathrm{cov}(q_T, q_N) + (\overline{q_T} - \overline{q_N})^2 && (5.27)\end{aligned}$$

Here, q_T and q_N are the analytical and numerical solutions of the advected field, respectively, M is the number of grid points, and σ is the variance. According to Eq. (5.25), the total error can be decomposed into E_{disp} and E_{diss} as

$$\begin{aligned} E_{diss} &= [\sigma(q_T) - \sigma(q_N)]^2 + (\overline{q_T} - \overline{q_N})^2 && (5.28)\\ \text{and } E_{disp} &= 2(1-\tau)\cdot\sigma(q_T)\cdot\sigma(q_N) && (5.29)\end{aligned}$$

where τ is the correlation coefficient between q_T and q_N. The three quantities E_{tot}, E_{diss}, and E_{disp} are calculated at each time step for each tagging methodology. Note that the analytical solution is only available when the flow field is uniform, hence these error measures are only available in the uniform wind field experiments.

The non-uniform wind field experiments are designed for evaluating the mass-conservative behaviour of the different tagging algorithms. As a measure of conservation of mass, at each time step the total sum of WV density[1] is calculated for the tagged moisture fields. While for the PTW and STW methods these sums can be calculated directly, the PTQ and q fields first have to be converted to WV density using Eq. (5.11). The temporal evolution of the total WV densities provide then information about the conservation of mass.

5.3 Tracer advection experiments

The basic characteristics of the applied numerical schemes are fundamental to understanding the differences between the results for the three tagging algorithms, and have already briefly been reviewed in Section 5.1.5. Here, first advection experiments for different initialisations and algorithms with a uniform wind field are presented. Then, the experiments with an oscillating flow field are discussed, in particular with respect to the conservation of mass.

5.3.1 1D-Tagging experiments using a uniform wind field

The three tracer algorithms PTQ, PTW, and STW are first compared within a model setup with a uniform wind field. After 100 integrations, a tracer initialisation of Type I (Fig. 5.5a) is advected about 50 grid points to the right (Fig. 5.6a). In comparison to the ideal solution, the PTQ tracer shows non-physical oscillations spreading to the left, as well as ripples behind the main anomaly. These negative values of the tagged WV density are physically meaningless. The amplitude of the initial block of tagged water is however well conserved. When in addition numerical diffusion is applied (PTQ_D), then the short wave anomalies are strongly damped, while longer wave length anomalies upstream of the tagged water block remain unchanged. The PTW and STW tagging methods with leap-frog flux form numerics and numerical diffusion (Fig. 5.6b) show slight differences. While the $PTW\text{-}LF_D$ implementation show the same upstream ripples as the PTQ and PTQ_D method, these have been suppressed in the $STW\text{-}LF_D$ case due to the imposed upper and lower boundaries to f (see Section 5.1.6). However, this corrective measure has also lead to a slightly larger phase error for the $STW\text{-}LF_D$ algorithm.

Comparing the PTW and STW methods for the Lax-Wendroff numerics, the difference with respect to overshoots and undershoots becomes even more apparent (Fig. 5.6c). The upper and lower limits imposed on f for the STW-LW method avoid overshoots and upstream ripples, however at the expense of conservation of mass (see below). With MPDATA numerics, the advection results of the PTW and STW numer-

[1]Water vapour density can also be considered as specific water vapour mass.

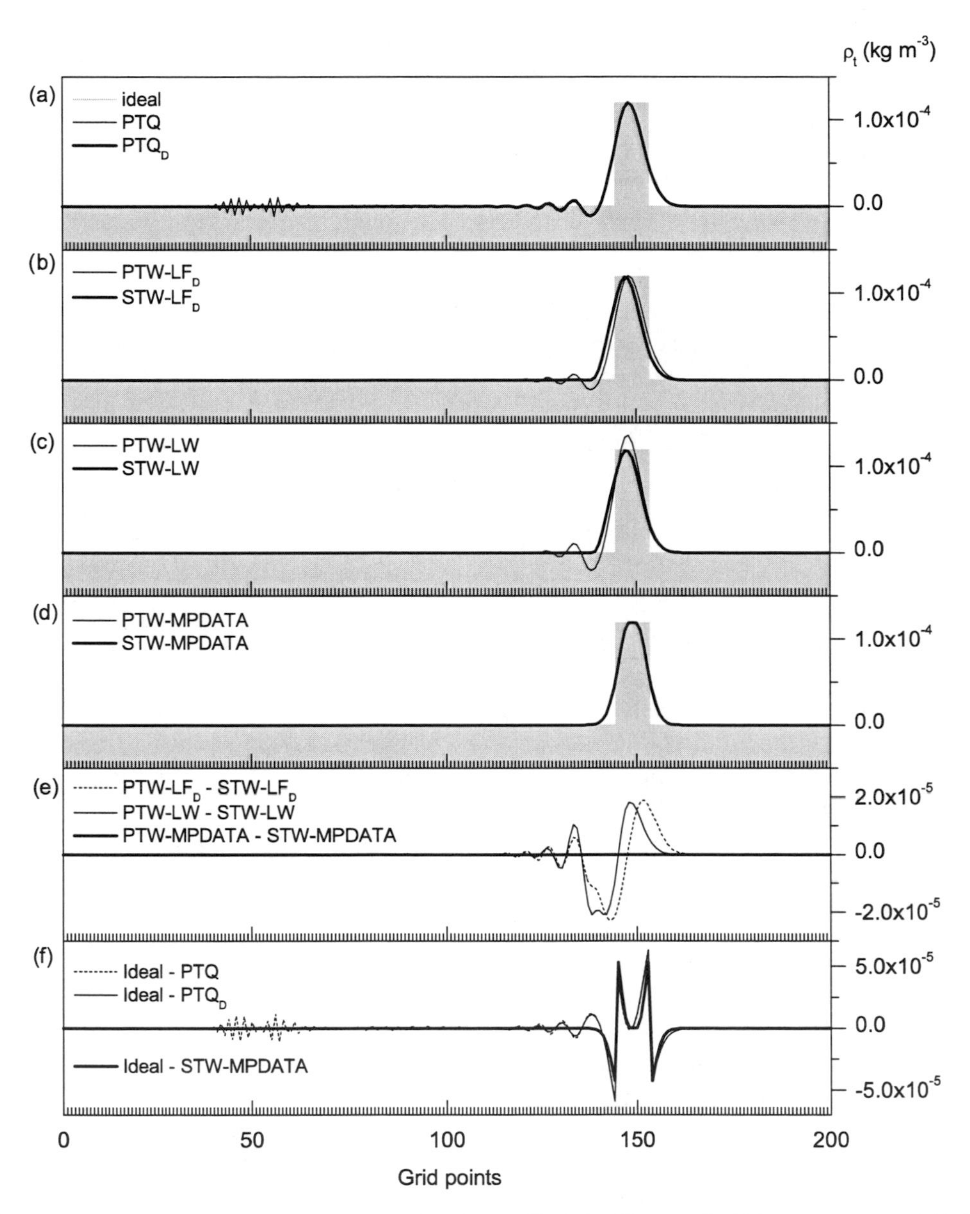

Figure 5.6: Comparison of the tagging algorithms PTQ, PTW, and STW for a rectangle initialisation after 100 integrations. Subscript D denotes application of numerical diffusion. LF and LW are the leap frog and Lax-Wendroff discretisation schemes, respectively.

ics are identical (Fig. 5.6d). The positive definite and non-oscillatory behaviour of the MPDATA scheme does not require to impose limits on f. Besides some diffusion at the edges, the advected anomaly is well preserved for both, the PTW-MPDATA and STW-MPDATA algorithms. No numerical diffusion is applied with the LF and MPDATA numerics, as the effect is either negligible or even negative (see Section 5.1.5).

The differences between different numerics for the PTW and STW tagging methods highlight the effects caused by imposing upper and lower limits to f (Fig. 5.6e). The LF and LW numerics result in differences up to 17%, and with an obvious phase shift. With MPDATA numerics, the PTW and STW methods are identical. Comparing the differences between the ideal solution and the numerical tracers (Fig. 5.6f), due the inherent diffusion of the numerical schemes, the strongest differences are visible near the edges of the initial anomaly. The STW-MPDATA method has a smaller phase error than both the PTQ and PTQ_D methods. Again, the selective damping of short wavelengths by explicit numerical diffusion is clearly visible.

The advection of a Type II initialisation (Fig. 5.5b) with the PTQ method produces a prominent short wave anomaly travelling upstream, that even is of larger magnitude than the remaining long wave anomaly (Fig. 5.7a). The short wave feature is almost completely filtered out when the explicit diffusion is applied. The PTW-LF_D and STW-LF_D methods show a larger difference than for the Type I initialisation (Fig. 5.7b). The correction of undershoots in the STW method has led to a significant increase in the marked water mass, and superimposed ripples upstream. For the LW numerics, differences are smaller than for the LF numerics, but also compared to the Type I initialisation, as no overshoots (only undershoots) have to be corrected by the STW method (Fig. 5.7c). For the MPDATA numerics, the PTW and STW methods again produce the identical result. Notably, due to implicit numerical diffusion, after a few time steps the amplitude of the initialisation is reduced by 1/2 for all algorithms considered.

Temporal evolution of error measures

The temporal evolution of the total error E_{tot} for a Type I initialisation and the PTQ, PTQ_D method peaks after 10 and ~30 time steps (Fig. 5.8a). Due to the removal of short-wave anomalies, the explicit diffusion significantly reduces the total error for this method. Both curves are reproduced in Fig. 5.8b as a reference. The PTW-LF method has the same error as the PTQ method, the same holds for PTW-LF_D (not shown). The error of the PTW-LW method increases gradually, and after about 70 integrations exceeds even that of the PTQ method. The total error is smallest for the PTW-MPDATA, and shows a monotonic increase. For the STW method, using LF numerics creates the largest total error (Fig. 5.8b). The same applies to the STW-LW implementation. Both strong increases in total error can be explained by the limits imposed on overshoots and undershoots, which cause a continuous growth of the initial anomaly. The total error of the STW-MPDATA method equals that of the PTW-MPDATA implementation.

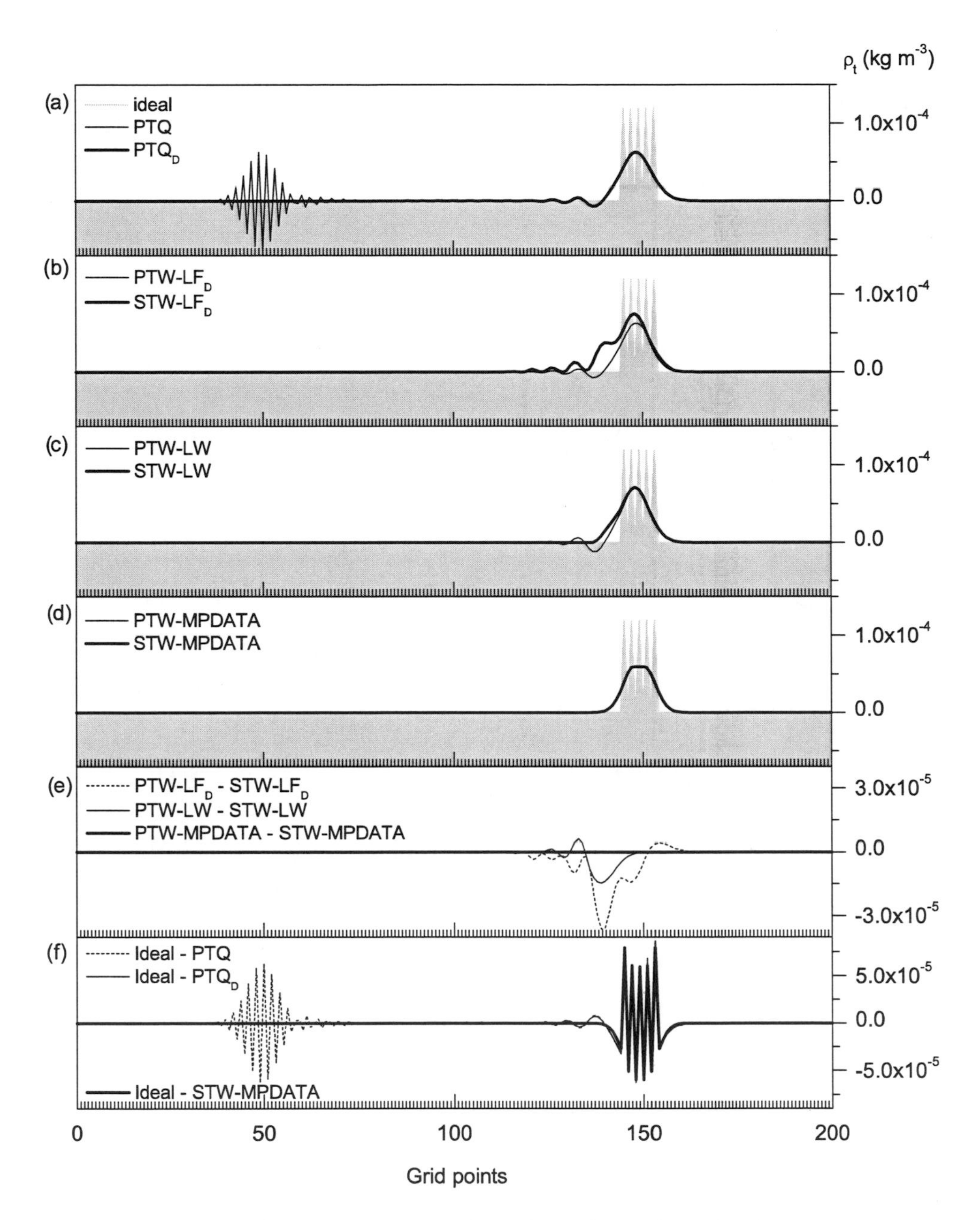

Figure 5.7: As Fig. 5.6, but for a Type II (zigzag) initialisation.

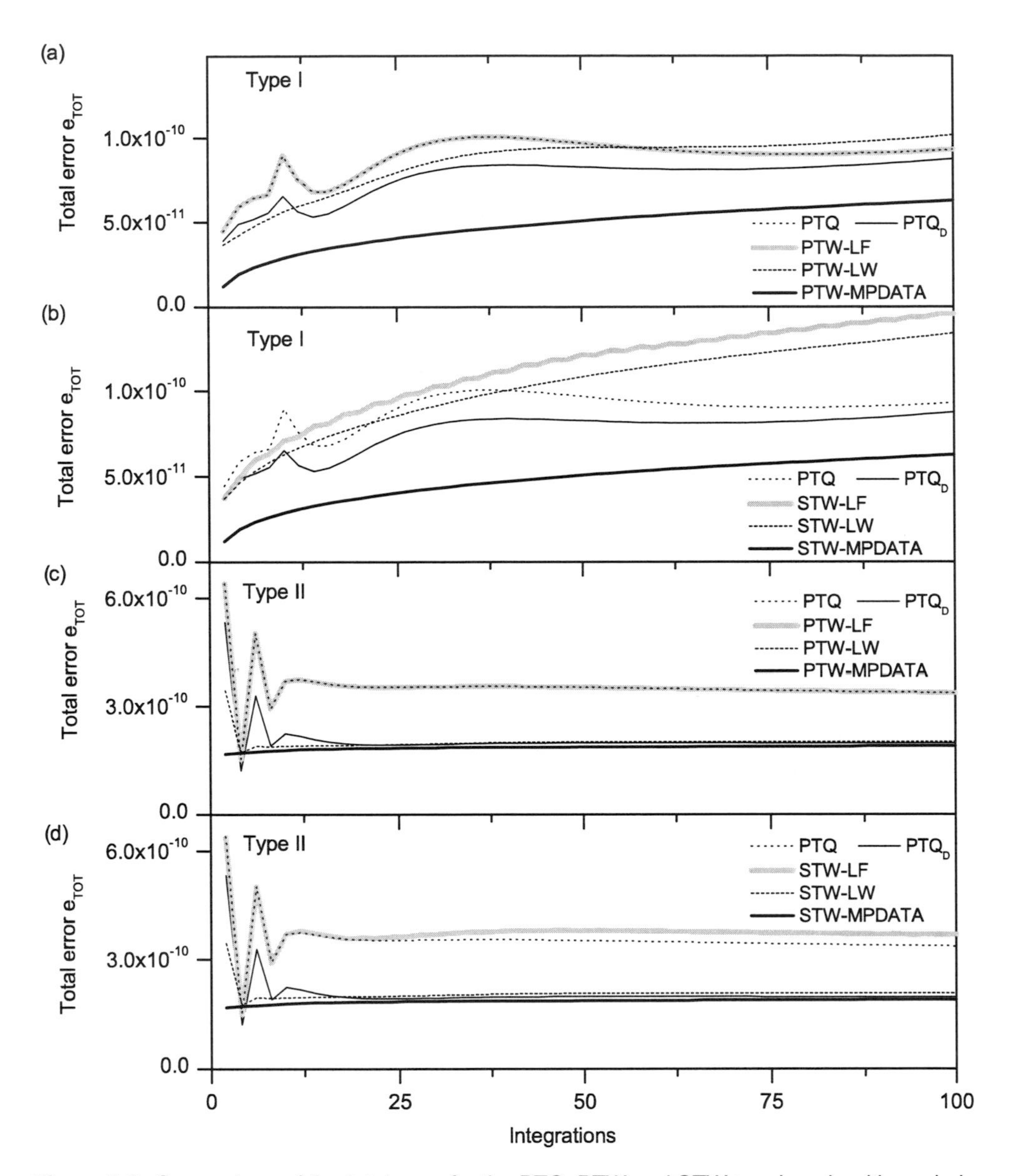

Figure 5.8: Comparison of the total error for the PTQ, PTW and STW tagging algorithms during advection of a Type I and a Type II initialisation for 100 time steps.

The total error for a Type II initialisation is largest during the 10 initial integrations, and remains fairly constant thereafter (Fig. 5.8c, d). The total error for leap frog numerics without explicit diffusion are offset to the other curves, pointing towards the short-wave anomaly upstream as the reason. The STW-LF and STW-LW methods again have larger total errors than their PTW counterparts. The PTW-MPDATA and

STW-MPDATA both form the same lower boundary to the total error evolution. From the experiments above, the MPDATA numerics appears as by far the most powerful choice for the PTW and STW tagging methods, and the LF and LW discretisations will not be considered further for these methods.

A more detailed examination of the temporal evolution of the various error measures given in Section 5.2.2 provides detailed insight into the dispersive and dissipative characteristics of the explicit numerical diffusion (Fig. 5.9). As noted above, for the PTQ method, the total error is strongly reduced with explicit diffusion. This is almost completely due to a reduced dispersion error, and only a slight increase in the dissipation error (Fig. 5.9b, c). Using explicit diffusion with the PTW-MPDATA method shows no change in the total error, which is however a compensational effect of an increased dispersion error and a decreased dissipation error. For the STW-MPDATA method, besides the dispersion error, the dissipation error increases due to overshoot correction, and accordingly the overall error increases as well.

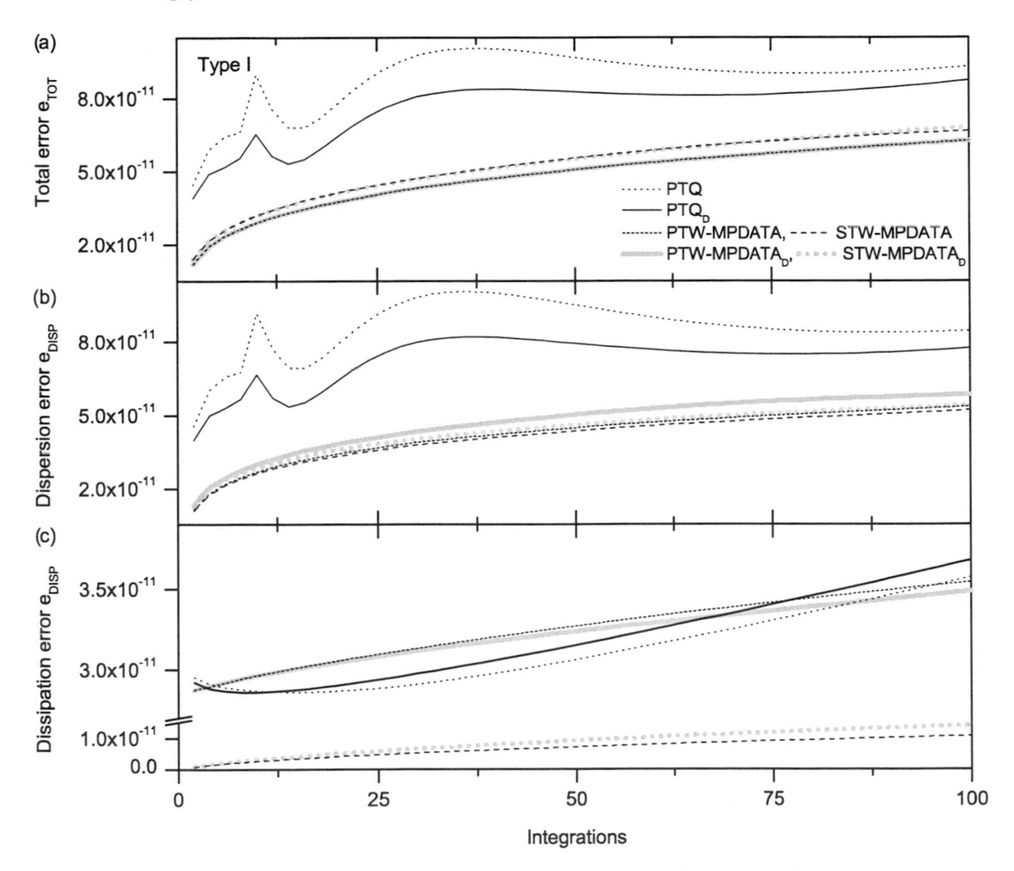

Figure 5.9: Temporal evolution of total error, dispersion error, and dissipation error for Type I initialisation. Both PTW and STW use MPDATA numerics. Note the enlarged scaling for the dissipation error.

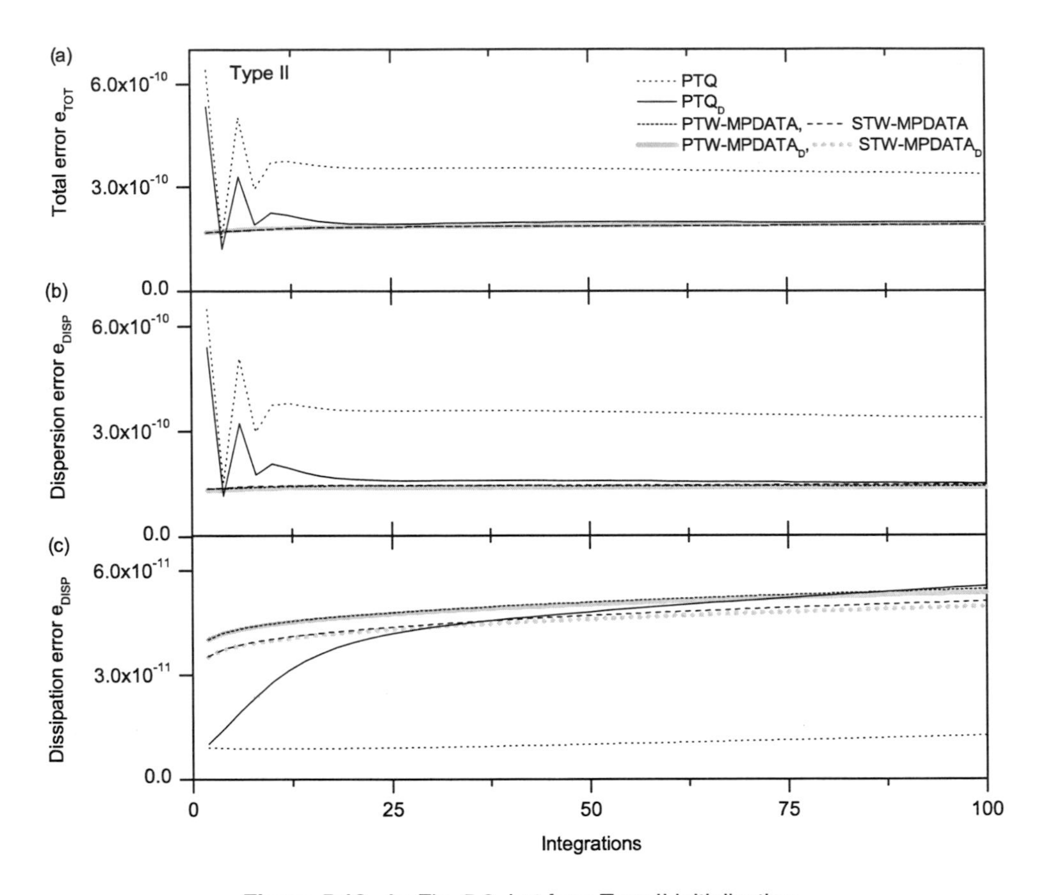

Figure 5.10: As Fig. 5.9, but for a Type II initialisation.

For a Type II initialisation, the changes due to explicit diffusion are less clear (Fig. 5.10). The temporal evolution of the error measures begins with a pronounced decrease, and remains at a constant value thereafter. The decrease in total error occurs during the smoothing of the initially pulsed tracer signal. While the dissipation error shows a small increase in the beginning, the overall decrease is almost completely due to the decreasing dispersion error. As for the Type II initialisation no overshoots due to explicit diffusion occur, the difference between the PTW and STW tracer signals is rather small during the integration period. The total error for the PTW-MPDATA and STW-MPDATA methods with and without diffusion remains virtually unchanged. Because of the only marginal changes to the long-term evolution of the error measures and even some increases in total error for the STW method, explicit diffusion should not invariably be applied for the PTW and STW tagging methods.

In summary, the first set of experiments demonstrates the ability of both the PTW-MPDATA and STW-MPDATA tagging approaches to serve as a qualitative tracer for advection with a uniform wind field. The characteristics of the advected tracer field are largely determined by the chosen discretisation, and are quantified by the error measures described in Section 5.2.2. Explicit numerical diffusion improves the LF advection, but leads to worse results for the MPDATA advection scheme. The PTQ tracer, as well as LF and LW numerics with the PTW and STW methods, prove less suitable for tagging applications due to numerical ripples, and the production of 'negative mass'.

5.3.2 1D-Tagging experiments using an oscillating wind field

A second set of 1D experiments is carried out using a sinusodial wind field (Fig. 5.4). In this setup, wind velocities vary sinusodially between 1.5 and 2.0 m s^{-1}. The Courant number accordingly ranges between 0.38 and 0.50, which guarantees the stability of the applied numerical schemes. The purpose of this setup is to test the different tagging algorithms for their mass conservative behaviour. For this setup, an analytical solution to compare with is not readily available, therefore no error measures as for the uniform wind field can be given. Instead, the temporal evolution of the total mass is considered.

The total specific mass (or WV total density) of the main WV field q, calculated with the CHRM LF numerics, is constant for the Type I and II initialisation (Fig.5.11a). With a Type III initialisation however, the total mass oscillates due to the non-conservative formulation of specific humidity advection in the CHRM. For the first two initialisations, the main WV field retains a constant total mass as the gradients are zero in the uniform specific humidity field. For the Type III initialisation however, non-zero gradients lead to changes in the main WV field during advection according to Eq. (5.7), which is not mass-conservative. The same applies to the PTQ$_{\mathrm{D}}$ method (Fig. 5.11b): Here, for all three initialisations non-zero gradients exist, and the non-conservative behaviour of the CHRM numerics becomes evident.

The PTW-MPDATA tagging method shows a fully conservative behaviour for all three initialisations for the oscillating wind field (Fig. 5.11c). The PTW-MPDATA method hence meets the important requirement of conservation of mass very well. The STW algorithm shows again a different behaviour for the three initialisations (Fig. 5.11d). While the Type II and III initialisations are advected conservatively, a decrease in mass is observed during the first 80 time steps during the advection of a Type I initialisation. This decrease occurs synchronously with the decrease in tagged water mass for a Type I initialisation. The decrease in mass is due to the synchronisation of the tagged WV field with the main WV in the STW-MPDATA method. This particular behaviour will now be further elucidated.

Following the temporal evolution of the advection of a block of marked water illustrates the differences between parallel and synchronised tagging methods. In a first setup, a Type I initialisation is placed at grid point 25–34 (Fig. 5.12). In this section of the model domain, the wind field is convergent, and hence the advected anomaly will be compressed. While the main WV field remains unchanged in this setup due to $\nabla q = 0$, the PTQ$_{\mathrm{D}}$ method results in an anomaly with the typical overshoots and undershoots.

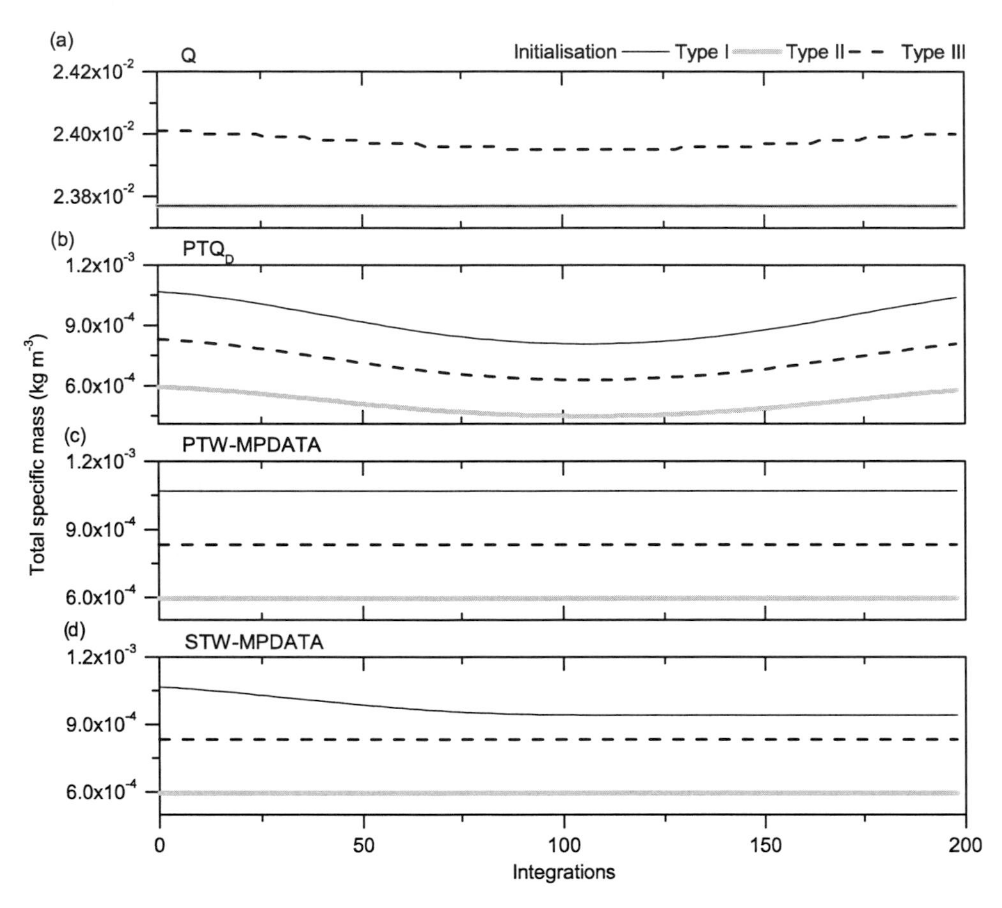

Figure 5.11: Temporal evolution of the total mass of the main WV field and the three tracer algorithms for a sinusodial wind field.

Accordingly, the PTQ_D advection result exceeds the main WV field at $t = 50$ (Fig. 5.12a, arrow). The same applies to the PTW-MPDATA method. As mass is conserved for this tagging method, the overshoots occur at $t = 50, 100, 150$ (Fig. 5.12b, arrows). Therefore, tagged water exceeds the main WV field at that grid cell. The STW-MPDATA method avoids these overshoots due to the synchronisation of the tagged WV field with the main WV field, however at the expense of violating the conservation of mass.

The experiment is repeated with a Type I initialisation placed at grid point 75-84, a region where the wind field is divergent (Fig. 5.13). The PTQ_D method shows again the typical overshoots and undershoots, hence the tagged water mass at these grid cells is larger than that of the main WV field (Fig. 5.13a). The PTW-MPDATA method now only shows one overshoot at $t = 200$, as implicit diffusion and the divergence of the wind field have dissipated the anomaly strongly enough so that the overshoot only occurs at the end of the next divergent region of the wind field (Fig. 5.13b). The STW-MPDATA method in this setup only adjusts the tagged water mass at time step $t = 200$ (Fig. 5.13c).

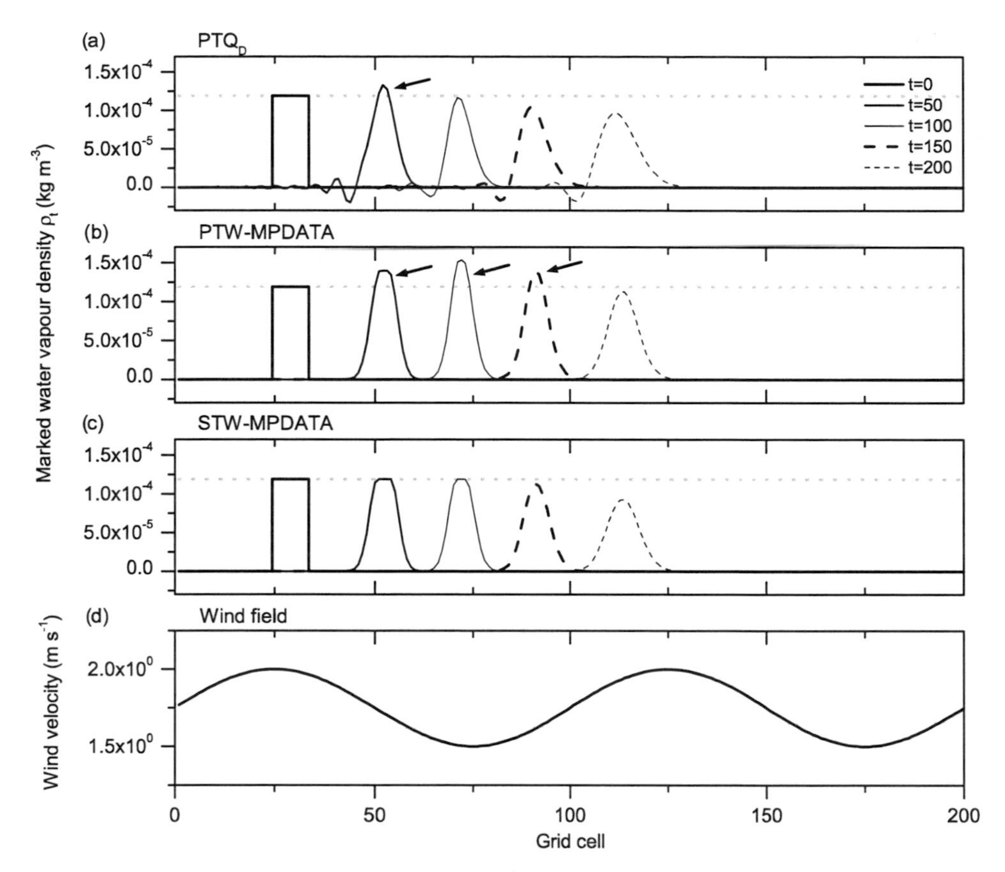

Figure 5.12: Temporal evolution of the advection of a Type I initialisation at grid point 25–34 for an inhomogeneous wind field with three tagging methods PTQ_D, PTW-MPDATA, and STW-MPDATA. The dotted gray line indicates the specific mass of the main WV field. Arrows denote overshoots of the tagged water w.r.t the main WV field.

A direct comparison of the total specific mass for the two setups of a Type I initialisation with the PTQ_D, PTW-MPDATA, and STW-MPDATA methods shows the intermediate position of the synchronised tracer approach (Fig. 5.14). In the case of an initialisation at grid point 25, the PTQ_D method shows first a decrease and then an increase in total tagged water mass, the PTW-MPDATA remains constant throughout the 200 integrations (Fig. 5.14a). The STW-MPDATA method takes an intermediate position, loosing some mass due to the synchronisation with the main field, and then remaining constant for the rest of the advection time. In contrast, for the initialisation at grid point 75, the PTQ_D method shows first an increase an then decreases again, while both the PTW-MPDATA and STW-MPDATA remain at a constant total specific mass. Only for the last 15 integrations, a slight loss in total specific mass can be seen for the STW-MPDATA method, again due to the synchronisation with the main field.

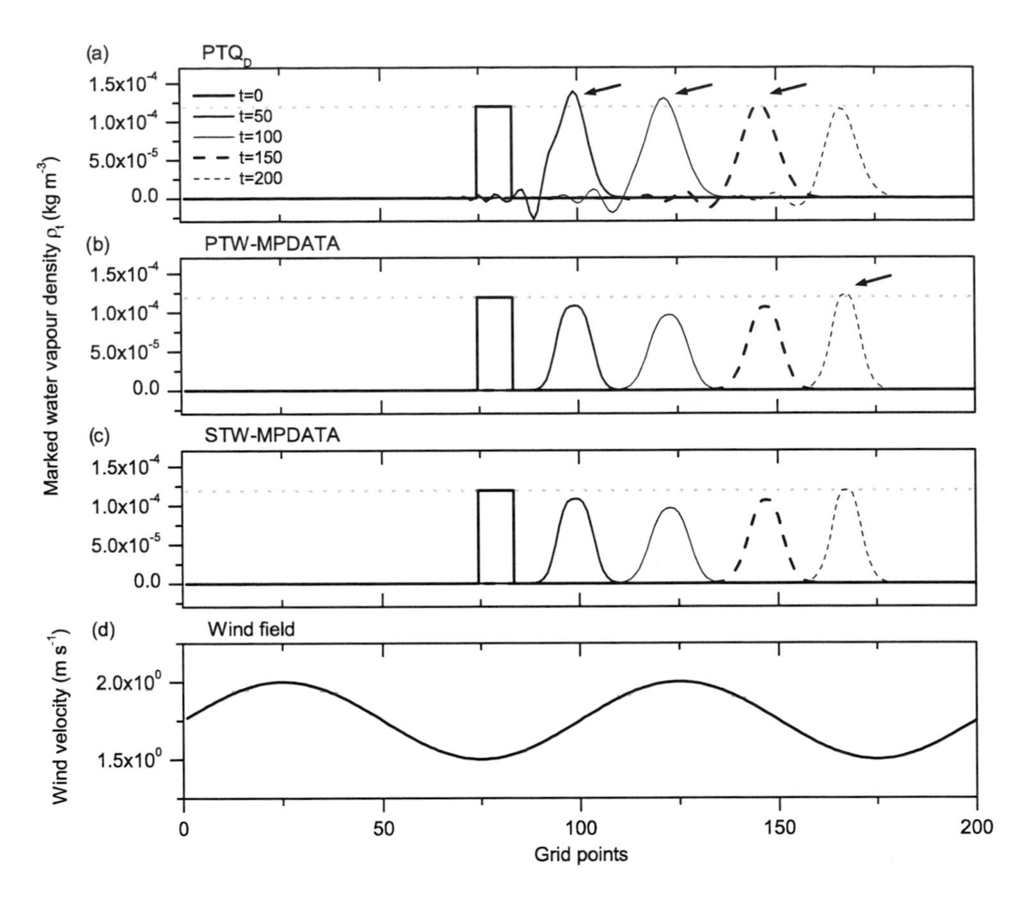

Figure 5.13: As Fig. 5.12, but for a Type I initialisation at grid point 75–84.

5.4 Concluding assessment of the tagging algorithms

Three tagging implementation strategies have been presented, and experiments conducted to test the behaviour during advection in a 1D environment. The findings can be summarised as follows:

PTQ: The parallel tracer with specific humidity advection uses the same discretisation scheme as the CHRM model for its moisture field advection. The apparent advantage of numerical consistency is however not justified: Due to steeper gradients in the tracer field, overshoots and undershoots occur, which do not occur in the main WV field. In addition, undershoots represent physically meaningless negative mass. Finally, for an inhomogeneous wind field, the tracer mass is not conserved with this tagging implementation. As these properties contradict the requirements for tagging methods set out in the beginning of this chapter, the PTQ method is considered as unsuitable for the tagging applications intended here.

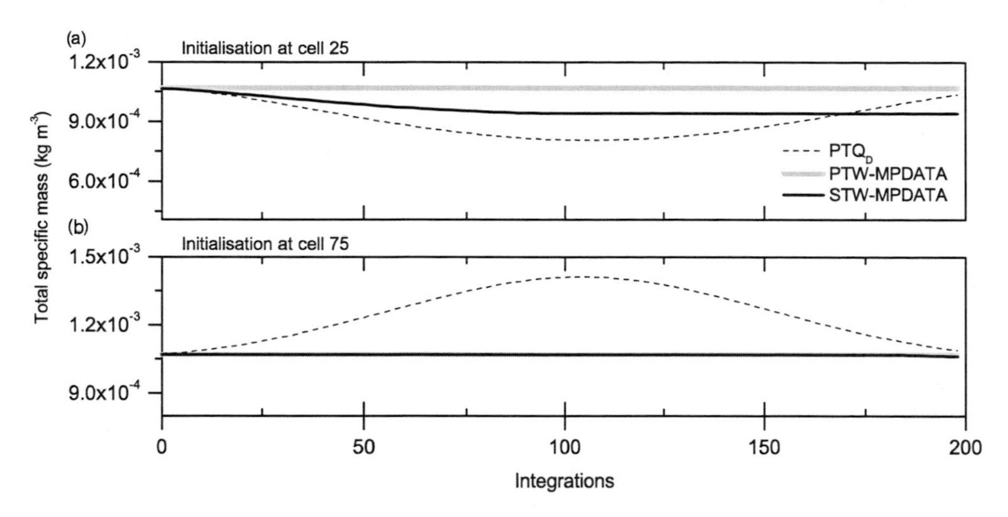

Figure 5.14: Temporal evolution of the total mass of the main WV field and the three tracer algorithms for a sinusodial wind field.

PTW: The parallel tracer with water mass advection uses the advanced MPDATA advection scheme. Instead of specific humidity, specific water mass is advected. The scheme has a small amplitude error, is mass conservative, and able to prevent the occurrence of negative values. However, as the main WV field and the tracer field are advected independently with different numerics, inconsistencies can evolve, such as cells containing more tagged water than is predicted from the advection of the main WV field. Thus, the advantages of the MPDATA scheme make the PTW method an option which could be considered for tagging applications, but is limited by a number of problems which could arise from inconsistencies between tagged and total water mass.

STW: The synchronised tracer with water mass advection uses again the MPDATA scheme with its advantageous properties. For a homogeneous wind field, the PTW and STW tagging methods are identical. As the STW method is synchronised with the main WV field, the tagged water mass can never exceed the total water mass at a grid cell. However, due to the main WV field numerics being non-conservative, the STW method can loose tagged water mass during synchronisation. This property of the STW method is both a limitation and a feature, ensuring a high degree of consistency between the tracer and the main WV field.

In conclusion, the STW approach takes an intermediate position between the PTQ and PTW methods. With respect to the intended applications, based on the assessment of the three tagging methods PTQ, PTW, and STW presented above, the STW method is chosen as the best implementation strategy for the CHRM model. The technicalities of the implementation in the CHRM model environment, as well as the treatment of tagged water fluxes from parameterisations, will be presented in the next chapter.

Chapter 6

Implementation of WVT in the CHRM Model

Implementing a water vapour tagging (WVT) method into a regional climate model (RCM) is an endeavour that touches upon most model components to at least some degree. In contrast to chemical tracers, such as ozone, water is an integral component of numerical weather prediction (NWP) models, that via diabatic processes feeds back into the evolution of the atmospheric flow. The particular problems of WVT in a NWP model is then that (i) the model behaviour must remain unchanged when tracers are released, while (ii) the large-scale and sub-grid scale processes in the model must redistribute the tracer in the same way as all water in the model is treated. This chapter details how these two problems are solved in the present WVT implementation.

All tagging implementations so far have been carried out in general circulation models (GCMs). Since in this work a tagging is implemented in a RCM, it is well worth to highlight some important differences between the two model types. The most obvious difference between a GCM and a RCM is the difference in spatial and temporal resolution. Current GCMs typically apply horizontal grid sizes between several to one degree, while RCMs can now resolve scales up to a few kilometres (and below). Naturally, the increased resolution is accompanied by a more detailed representation of the orography and the land cover. Due to the increased resolution, parametrisations of sub-grid scale processes are more detailed in RCMs, for example with respect to phase change processes of water in the atmosphere (cloud microphysics) or the soil (plant cover, soil storage). A RCM tagging implementation will therefore require a more detailed coverage of these parametrised processes.

All sub-grid scale processes which affect water vapour in the model are represented by parametrisations. There is no commonly accepted way to treat tracers in parametrisations. Numaguti (1999) assumes an equilibrium between water vapour and cloud water at each grid cell, an assumption that may be fulfilled in GCMs but not in RCMs. Soil processes are typically only represented in a rather crude manner, which may bias precipitation recycling studies towards faster or stronger precipitation recycling.

Bosilovich and Schubert (2002) applied a bulk approach, where for all tracers the tendencies due to parametrisations are treated as proportional to the tendencies of the total moisture field at each grid cell. This however neglects vertical redistribution due to precipitation re-evaporation and convective updraughts and downdraughts. This could possibly lead to the systematic biases observed in the equatorial regions observed by Bosilovich and Schubert (2002). Hence, a different, more detailed treatment of parametrised processes will be adopted here.

Following a short general description of the CHRM (Climate High Resolution Model), the description of the implementation will first focus on aspects related to the setup and output of a tagging simulation. Then, the implementation of tracer advection and diffusion are described, followed by a treatment of WVT in the various parametrisations related to moisture transport and phase changes in the model.

6.1 The CHRM model

The CHRM stems from the German Weather Service's (DWD) GM/EM/DM model chain (Majewski 1991). Originally developed for operational weather forecasts, the EM/DM has later been extended to run seasonal to multi-annual simulations (Schär et al. 1999). In particular, a more sophisticated soil model was implemented (Vidale et al. 2002). The special suitability for climate simulations, which is reflected in the name change to CHRM, is one important reason to rely on this particular model version for the tagging implementation.

Another important reason is the long record of realistic climate simulations and sensitivity studies with the EM/CHRM model. This encompasses precipitation validation studies (Lüthi et al. 1996; Schmidli et al. 2002; Frei et al. 2003), flood-related processes (Keil et al. 1999), and a regional climate model intercomparison (Mladek et al. 2000). A particular focus has been on studies of the soil moisture feedback (Schär et al. 1999; Heck et al. 2001; Vidale et al. 2002; Seneviratne et al. 2002), also with regard to predictability (Fukutome et al. 2001; Vidale et al. 2003). The model has also been applied to study extreme events and future climate change (Schär et al. 2004).

Technically, the CHRM operates in a rotated coordinate system with a typical (non-uniform) grid resolution of 14-50 km. In the horizontal, an Arakawa-C grid is used for the spatial discretisation. In the vertical, a hybrid sigma/pressure co-ordinate system (η co-ordinate) is applied (Simmons and Burridge 1981). The advection numerics use a leap-frog discretisation with an Asselin filter (coefficient $\gamma = 0.15$) for the prognostic variables momentum (u, v), temperature (T), water vapour (q_v), and cloud water (q_c). Vertical advection and diffusion (including turbulence and surface-layer flux parametrisation) are integrated with an implicit leap-frog scheme. In the horizontal, a 4th-order explicit diffusion is applied to u, v, T and q_v. In contrast to the EM/DM models, there is no option for a semi-Lagrangian advection scheme in the CHRM. A semi-implicit gravity wave correction (Temperton 1988) is applied after each time step. Grid-scale precipitation is parametrised with a bulk microphysics scheme following Kessler (1969) and Lin et al. (1983). For convective precipitation, the Tiedtke (1989) mass-flux convection scheme is used. The soil model is of Dickinson (1984) type, with 3 vertical layers (Vidale et al. 2002). The model domain is initialised with an implicit normal-mode initialisation (Temperton 1988). In the horizontal, the Davies (1976) boundary relaxation with an 8 point wide boundary zone is used. In the vertical, a radiative upper boundary condition is applied at the top of the model domain.

6.1.1 Code organisation and changes

The CHRM model code in its present version (2.3 WLWC, Dec. 2001) is spread over ~120 Fortran90 files. A short overview of the dependencies of the code sections is given here to clarify (i) the order in which certain routines are called, and (ii) where code changes were necessary for the implementation of the WVT. In Table 6.1, calling hierarchies are shown by indentation of the routine names. Some less relevant routines have been left out for the sake of clarity.

Input parameters for the tagging are read in the routines `readnlst` and `wvt_slab`. Grid-scale and convective precipitation tagging are covered in the routines `pargsp` and `parcon` and their subroutines. The advection and diffusion of the tracer takes place in `progexp`. Condensation processes are handled in `condens`. Boundary relaxation is performed in `lb_relax`. Finally, output tracer variables are written to GRIB files in `pp_org` and the related subroutines (see Sec. 6.1.4). All changes to the code related to the WVT implementation are bracketed by introducing (`WVT code`) and trailing (`End WVT code`) statements.

6.1.2 Tracer initialisations

Numaguti (1999) defined four different basic types of WVT experiments. The first and (most intuitive) approach (I) is to assign a tag to a water 'molecule' as it evaporates from a particular source area. This tag is only lost when the tagged water is removed from the atmosphere as precipitation. A second approach (II) is to assign more 'sticky' origin tag to water molecules. These tags are also kept when water is precipitated to the land surface, and then re-evaporated from the surface to the atmosphere. This tagging experiment identifies the most recent ocean evaporation source of a WVT. A third tracer experiment (III) introduces the tracer age by incrementing the tracer fraction (initially 0) after each time step. In combination with experiments I and II, the additional age tracer can then be used to identify the duration of atmospheric transport, or the time since evaporation from an ocean surface. The fourth tracer experiment (IV) counts the number of re-evaporations of a tracer, increasing from zero at evaporation from the ocean surface by one for each re-evaporation. A fifth tracer experiment (V), that was only vaguely indicated by Numaguti (1999), and is further developed here, is to assign

Table 6.1: Organisation of the CHRM core code. Changed routines are denoted by a †, new routines added for the WVT are printed in bold face.

Routine	Description
hrmorg	*CHRM main routine*
readnlst†	*read input from file INPUT_HRM*
allocatefields†	*allocate all memory*
wvt_init	*initialise atmospheric WVTs*
wvt_slab_init	*initialise WVT surface slab*
progorg†	*model integration loop*
physics†	*physics parametrisations*
pargsp†	*grid-scale precipitation*
partura	*atmospheric turbulence parametrisation*
parturs	*surface turbulence parametrisation*
pardickin	*Dickinson soil model*
(parsoil_1)	*pre-Dickinson soil model part I, obsolete*
moist_c	*prepare moist convection*
parcon†	*convection parametrisation*
parsoil_2	*soil model part II*
...	
prog_s_03	
progexp†	*advection/diffusion of prognostic variables*
wvt_mpdata3d	*WVT advection*
condens†	*added call to condens instead of inline code*
prog_s_06	
condens†	*condensation after semi-implicit corrections*
prog_s_07†	
asselin†	*Asselin filter of prognostic fields*
lb_relax†	*boundary relaxation*
condens†	*condensation after corrections*
prog_s_08†	
nearsfc†	*convert precipitation rates to precipitation*
wvt_sync	*synchronise q_t with background q field*
wvt_stats	*calculate and write WVT statistics*
pp_org†	*post-processing and GRIB output*
...	

a tag to all water mass that resides in or crosses a particular section of the atmosphere during a specified period of time.

Three different options are available to release the tracer in the CHRM model domain during a specific period of time: (i) Tagging of all water in a specified volume of the model domain, (ii) tagging of all water entering the model domain from a boundary, (iii) tagging of all water vapour evaporating from a specified surface area during the model integration. The first two options correspond to Numaguti (1999)'s experiment V, while the third option corresponds to his experiment I. Options (i) and (ii) are available via so-called *box initialisation*, while option (iii) is implemented as a so-called *slab initialisation*. Depending on the desired tagging application, these initialisation options can be used separately or in combination. After describing the general input syntax for the tagging scheme in CHRM, the following sections give examples for each of these initialisation types.

Namelist input

CHRM model runs are controlled via namelist input files. During startup, CHRM reads in all parameters from the file INPUT_HRM, which is located in the startup directory. Two new namelists were added to control the initialisation and calculation of WVTs. Only these two namelists (/wvt_ctrl/ and /wvt_box/) are described here, information on the standard namelist control parameters is given in DWD (1995).

The namelist /wvt_ctrl/ controls some general characteristics and partly the initialisation of the WVTs. The advection scheme which is used for the tracer advection is chosen with the parameter *wvtalg* (see namelist example below). Correspondingly, for the background field q_v, q_c, one can choose between leap-frog and the newly implemented MPDATA advection scheme (Smolarkiewicz 2005) with the parameter *qalg*. The whole tagging implementation can be switched on or off using *use_wvt*. Currently, only one WVT variable is available, hence the parameter *wvt_id* has no effect as yet. The parameter *boxes* determines the number of /wvt_box/ namelist boxes which are expected for box initialisation input (see /wvt_box/ below).

```
&wvt_ctrl
     qalg=0,               water vapour and cloud water advection algorithm
                           0: leap frog, 1: MPDATA
     use_wvt=.true.,       switch WVT on/off
     wvtalg=0,             WVT advection algorithm
                           0: upstream, 1: leap frog, 2: Lax-Wendroff, 3: MPDATA
     wvt_id=1,             index of the WVT (currently not used)
     boxes=1,              number of WVT initialisation boxes
/end
```

The namelist /wvt_box/ contains the parameters for a time-space section in the model atmosphere which is to be initialised with water vapour tracers (see examples below). All water vapour and cloud water within such a box is initialised as tracer during the respective time interval. CHRM coordinates run from 0 to *xmax* indices in west-east direction, from 0 to *ymax* indices in south-north direction, and from 0 to *zmax* levels from the top of the atmosphere to the surface. The time interval is given in seconds of model integration time. If required, individual boxes can be switched on or off with the parameter *use_box* instead of deleting the respective lines in the input file. For reference purposes, a box identification number may be assigned to the parameter *box_id*. The number of /wvt_box/ namelists in the file INPUT_HRM must match the parameter *boxes* in the namelist /wvt_ctrl/.

An example for the initialisation of a box in the upper troposphere and in the centre of the model domain at model time 0 (initialisation option (i)) is:

```
&wvt_box
    box_id=1,                    box identification number
    use_box=.true.,              switch box on/off
    xstart=30, xend= 35,         box start and end coordinates (x grid index)
    ystart=20, yend= 25,         box start and end coordinates (y grid index)
    zstart= 8, zend= 15,         box start and end coordinates (z grid index)
    tstart= 0, tend= 0,          box start and end in time (s)
/end
```

To initialise all tracer entering the domain along the southern boundary during the first hour of model time (initialisation option (ii)), the following box initialisation parameters could for example be used:

```
&wvt_box
    box_id=1,                    box identification number
    use_box=.true.,              switch box on/off
    xstart= 0, xend= 90,         box start and end coordinates (x grid index)
    ystart= 0, yend= 7,          box start and end coordinates (y grid index)
    zstart= 1, zend= 28,         box start and end coordinates (z grid index)
    tstart= 0, tend=3600,        box start and end in time (s)
/end
```

Slab initialisation

To release tracer via evaporation from a source volume, an additional slab initialisation file is specified via the parameter *slabfile* in the namelist /wvt_slab/. Slab initialisation can be switched on and off via the parameter *use_slab*:

```
&wvt_slab
    use_slab=.true.,                     surface slab initialisation on/off
    slabfile='slabs/slab_med',           path and name of slab initialisation file
/end
```

For each grid point at the model surface, the slab file lists the time intervals in seconds of model integration time during which this grid point acts as an evaporative source of the tracer. An example for the format of such a slab file is:

```
1256  1                    number of initialisation points, WVT id
1     1     1              x-index, y-index, number of time intervals
0     2600                 tracer release time start, end (s)
2     1     2              tracer release at two time intervals
0     2600  4000  8000     tracer release time start_1, end_1, start_2, end_2 (s)
3     1     1
0     2600
4     1     1
...
```

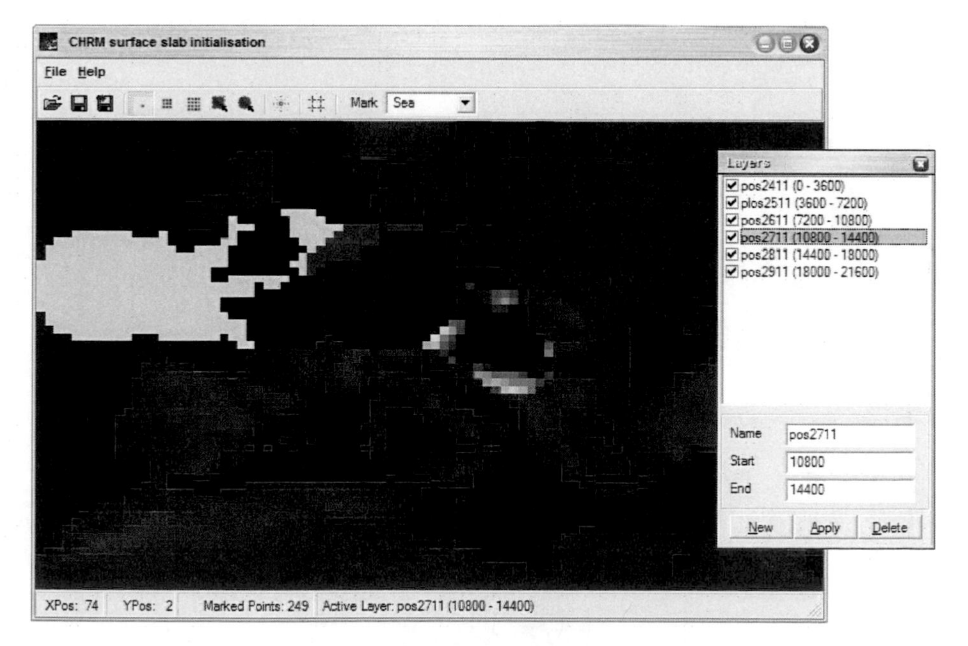

Figure 6.1: Snapshot from the slab initialisation tool *slab_init* for an initialisation area moving with the track of a cyclone. The main window shows a series of initialisation areas which are active during different time slices listed in the 'Layers' window.

In the example for the namelist /wvt_slab/, the slab file *slabs/slab_med* is specified which tags all water vapour evaporating from the Mediterranean during the model integration time (initialisation option (iii)). The advantage of such slab initialisation files is their flexibility in terms of the spatial domain that acts as a tracer source. Typical examples are either tracer release from certain ocean basins during different time intervals as in the example above, or a source area moving with the track of a cyclonic system. However, the difficulty is to specify such an input file efficiently. In order to simplify this initialisation task, the software tool *slab_init* has been created which allows to define tracer release areas and time intervals in a fashion similar to a typical painting software. Different layers of tracer release areas can have different release intervals assigned. A snapshot of this software tool shows a typical example with an active layer in red, and several other layers which are currently not edited in gray (Fig. 6.1). The software directly exports the initialisation data in the slab file format defined above. Copies of the software for Linux and Windows computers are available from the webpage `http://www.iac.ethz.ch/people/sharald`.

Table 6.2: WVT statistic output which is written to the file WVT_OUT. For each time step, one row which contains the 19 parameters from left to right is saved.

1. model integration time (s),
2. total water (kg m^{-3}),
3. total tagged water (kg m^{-3}),
4. % of tagged water relative to the total water,
5. total water vapour (kg m^{-3}),
6. total tagged water vapour (kg m^{-3}),
7. % of tagged water vapour relative to the total water vapour,
8. total cloud water (kg m^{-3}),
9. total tagged cloud water (kg m^{-3}),
10. % of tagged cloud water relative to the total cloud water,
11. accumulated total precipitation (mm),
12. accumulated total tagged precipitation (mm),
13. % of accumulated tagged precipitation relative to
    the total accumulated precipitation,
14. accumulated total grid-scale precipitation (mm),
15. accumulated total tagged grid-scale precipitation (mm),
16. % of accumulated tagged grid-scale precipitation relative to total,
17. accumulated total convective precipitation (mm),
18. accumulated total tagged convective precipitation (mm),
19. % of accumulated tagged convective precipitation relative to total.

6.1.3 WVT output

In tagging mode, CHRM keeps track of a number of quantities concerning the tagging implementation. At each time step this information is written to the output file WVT_OUT in the startup directory of CHRM. The calculation of the statistical information excludes the horizontal 8-point boundary zone. Therefore, the initialisation box in the 2nd example above does not affect the calculation of WVT statistics. The logged quantities are listed in Table 6.2.

6.1.4 GRIB output

Model output from the CHRM is written in the GRIB data format. Typical meteorological variables are standardised to some degree in so-called GRIB-tables, defined by the World Meteorological Organisation (WMO) or, in the case of CHRM output, the DWD. The tracer implementation required to declare the new table no. 204 for the definition of the tagging variables. Up to 256 variables, specified by the so-called GRIB code, can be defined in this table. Currently, 13 variables are defined in GRIB table 204 (see Table 6.3). CHRM writes any of these variables to the output files if they are specified in

Table 6.3: New GRIB variables implemented in CHRM for the WVT. $f_{x,t}$: fraction of tagged water vapour ($x = v$) or cloud water ($x = c$), $\Delta q^{y}_{x,t}$: tendency of tagged mixing ratio due to grid-scale precipitation (y =gsp) or convection (y =con), $R^{y}_{x,t}$: rate of tagged grid-scale or convective rain ($x = r$) or snow ($x = s$), $P^{y}_{x,t}$: accumulated precipitation total of tagged grid-scale or convective rain or snow.

Variable	Code	Table	Level[1]	SI unit	TRI [2]	R[3]	Note
WVT_QV	1	204	110	1	10	4	$f_{v,t}$
WVT_QC	2	204	110	1	10	4	$f_{c,t}$
DQV_WGSP	3	204	110	s^{-1}	10	3	$\Delta q^{gsp}_{v,t}$
DQC_WGSP	4	204	110	s^{-1}	10	3	$\Delta q^{gsp}_{c,t}$
DQV_WCON	5	204	110	s^{-1}	10	3	$\Delta q^{con}_{v,t}$
PRR_WGSP	6	204	1	$mm\,s^{-1}$	10	2	$R^{gsp}_{r,t}$
PRS_WGSP	7	204	1	$mm\,s^{-1}$	10	2	$R^{gsp}_{s,t}$
PRR_WCON	8	204	1	$mm\,s^{-1}$	10	2	$R^{con}_{r,t}$
PRS_WCON	9	204	1	$mm\,s^{-1}$	10	2	$R^{con}_{s,t}$
RAIN_WGSP	10	204	1	$kg\,m^{-2}$	4	2	$P^{gsp}_{r,t}$
SNOW_WGSP	11	204	1	$kg\,m^{-2}$	4	2	$P^{gsp}_{s,t}$
RAIN_WCON	12	204	1	$kg\,m^{-2}$	4	2	$P^{con}_{r,t}$
SNOW_WCON	13	204	1	$kg\,m^{-2}$	4	2	$P^{con}_{s,t}$

[1]Level 110: 3D atmospheric field, Level 1: 2D surface field
[2]Time range indicator (TRI) 10: instantaneous field, TRI 4: accumulated field.
[3]Range (R): Number of output dimensions of the field.

the file INPUT_HRM in the namelist /gribout/ in the parameter *yvarml* (for output on model levels) or *yvarpl* (for output on pressure levels). Further information on defining GRIB output for CHRM is available in the standard documentation (DWD 1995).

6.2 Large-scale dynamics

The numerics of the CHRM model integrate the prognostic equations for momentum, temperature, and moisture in the three-dimensional model domain. For the grid resolutions typically applied here (~14-50 km), tendencies due to large-scale advection are explicitly resolved, while a range of subgrid-scale processes, such as precipitation, have to be parametrised. This section describes the tagging implementation regarding all processes related to large-scale transport, while the tagging procedure concerning all processes involving phase changes is covered in Section 6.3.

Moisture transport in the CHRM is governed by the prognostic equation

$$\frac{\partial q}{\partial t}+\frac{1}{a\cos\varphi}\left(u\frac{\partial q}{\partial\lambda}+v\cos\varphi\frac{\partial q}{\partial\varphi}\right)+\dot{\eta}\frac{\partial q}{\partial\eta}=-\mathcal{F}_H^q-g\left(\frac{\partial p}{\partial\eta}\right)^{-1}\frac{\partial q}{\partial\eta}+\left(\frac{\partial q}{\partial t}\right)_{\text{sub}}-\mathcal{C}^q-\mu(q-q_R), \tag{6.1}$$

where q stands for the specific humidity of water vapour or cloud water, a is the radius of the earth, φ and λ are longitude and latitude of the rotated coordinates, u and v are the horizontal wind components, η is the vertical coordinate, $\mathcal{F}_H^q$ is the horizontal diffusion of moisture, g is the acceleration due to gravity, p denotes pressure, $\mathcal{C}^q$ is the tendency due to condensation, μ is the boundary relaxation parameter, and q_R is the boundary value. The l.h.s terms of Eq. (6.1) represent the tendencies due to horizontal and vertical advection in the rotated coordinate system (Sec. 6.2.1). The r.h.s terms from left to right stand for the moisture tendencies due to horizontal diffusion (Sec. 6.2.2), vertical diffusion (Sec. 6.2.3), sub-grid scale precipitation processes (Sec. 6.3.2–6.3.3), condensation (Sec. 6.3.1), and the prescribed boundary fields (Sec. 6.2.4), respectively.

The CHRM model numerics integrate Eq. (6.1) in several steps. First, the tendencies due to precipitation parametrisations and horizontal diffusion are calculated from time level $n-1$. Then, the horizontal advection tendency is calculated from an explicit leap-frog discretisation. Next, all tendencies are gathered in an equation system for each grid point, and including the vertical diffusion and advection terms solved with an implicit leap-frog method for time level $n+1$. Finally, condensation due to supersaturation is accounted for.

As demonstrated in Chapter 5, the synchronised tagged water (STW) algorithm in combination with the MPDATA advection scheme is a viable option for a tracer implementation in a model with a leap-frog advection scheme. However, the particular stepwise integration of Eq. (6.1) in the CHRM requires a careful implementation of the tagging methodology. Firstly, the spatial and temporal discretisation of the MPDATA advection scheme has to be matched with that of the CHRM. Second, the split horizontal and vertical advection have to be replaced by a single 3D advection step for MPDATA advection. Third, the horizontal diffusion, precipitation, and condensation terms have to be solved for the time level n instead of $n-1$. Finally, the implicit vertical diffusion has to be stripped from the vertical advection terms.

The integration of Eq. (6.1) is carried out in the routine `progexp` (see Table 6.1). The sequence of calculations is indicated in Table 6.4. As the tracer advection partly requires information from time level $n+1$ (see Sec. 6.2.1), but has to be carried out before the treatment of condensational processes, it was inserted after the completion of advection and before the condensation adjustment. Synchronisation of the tracer with the background field (see Chapter 5) is performed in the routine `wvt_sync` later (see Table 6.1).

6.2.1 Horizontal and vertical advection

For the horizontal discretisation, a staggered Arakawa-C grid is used in the CHRM model (Fig. 6.2a). Mass variables, such as q_v, q_c are placed at the cell centre (full levels) while flux variables (u, v) are located between cells (half levels). The horizontal staggering in the MPDATA routine is shifted by one grid index in the upstream direction (Fig. 6.2a). For use in MPDATA, horizontal advection velocities therefore have to be shifted accordingly. Close to the surface, the η vertical coordinate system has terrain-following levels, which gradually change into pressure levels with increasing altitude (Fig. 6.2b). The vertical staggering of the CHRM numerics and MPDATA are identical, and no shifting is required.

Table 6.4: Sequential calculation of the large-scale advection routine `progexp` in the CHRM. Changed or new sections due to the tagging implementation are printed in boldface. The tracer synchronisation is called outside of the routine `progexp`.

	Water transport process	Numerics
1.	Horizontal and vertical momentum advection	
2.	Horizontal moisture advection	explicit
3.	Horizontal moisture diffusion	explicit
4.	Vertical moisture advection and diffusion	implicit
5.	Filling of negative moisture values	
6.	**Tracer advection (3D MPDATA)**	explicit
7.	**Optional: Horizontal tracer diffusion**	explicit
8.	**Vertical tracer diffusion**	implicit
9.	**Filling of negative tracer values**	
10.	**Condensation**	
11.	**(Tracer synchronisation)**	

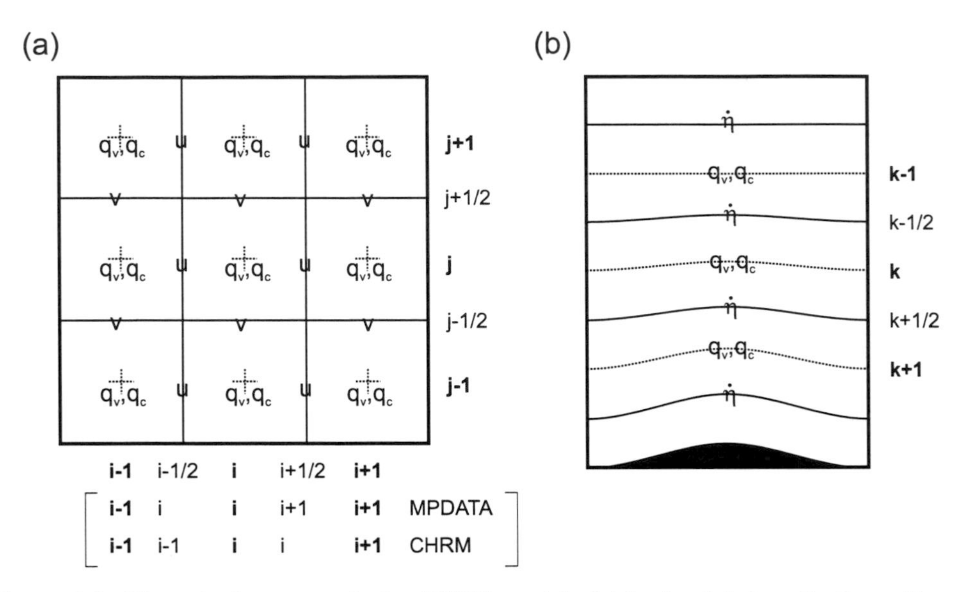

Figure 6.2: Discretisation space in the CHRM model. (a) horizontal discretisation with u, v at the staggered positions and q_v,q_c at the mass centres of the cells. The bracketed section shows the shifted model grid indices for CHRM and MPDATA. (b) Vertical discretisation in the η coordinate system with $\dot{\eta}$ at the half levels and q_v, q_c at the full levels.

The CHRM uses a leap-frog time discretisation for the whole dynamical core. In contrast, the MPDATA advection scheme uses a forward time step. The two temporal discretisations can coexist without difficulties as long as information from the correct time levels are fed into each advection scheme (see Chapter 5). In the CHRM, water vapour and cloud water are advected as specific humidities. Instead of using velocities for the advection, so-called mass flux velocities ($u \cdot \Delta p$) are used, which improve the conservation of mass. Still, as was shown in Chapter 5, it is difficult to conserve the mass of water with leap-frog advection. The MPDATA algorithm instead is specifically designed for the conservation of the advected density variables in arbitrary coordinate systems (Smolarkiewicz 2005). A call of the advection routine can be written as

$$\mathbf{G} \cdot \rho \cdot q^{n+1} = \text{MPDATA}(\mathbf{G} \cdot \rho \cdot q^n, \mathbf{v}_a^{n+1/2}) \tag{6.2}$$

where $\mathbf{G}$ is a matrix that contains coefficients which, similar to a Jacobi matrix, allow for arbitrary coordinate transformations, and ρ is the density of air. The temporal staggering of the advection velocity

$$\mathbf{v}_a^{n+1/2} = 0.5(\mathbf{v}_a^n + \mathbf{v}_a^{n+1}) \tag{6.3}$$

is required to maintain 2nd-order accuracy of the advection scheme (Smolarkiewicz 2005). The advection velocities are handed over to the MPDATA routine as advective Courant numbers:

$$\mathbf{v}_a^n = \begin{pmatrix} u^n \frac{\Delta t}{\Delta \lambda}, \\ v^n \frac{\Delta t}{\Delta \phi}, \\ \dot{\eta}^n \frac{\Delta t}{\Delta p^n} \end{pmatrix} . \tag{6.4}$$

The horizontal wind velocities can be transferred to the model coordinate system according to Eq. (6.1). In order to remain as numerically consistent with the CHRM numerics as possible, the same 'mass flux' velocities ($\mathbf{v}_a \cdot \Delta p$) are used for the MPDATA advection. This simplifies the handling of the vertical coordinate system. The components of matrix $\mathbf{G}$ are therefore set to the thickness of the cells Δp. The advected quantity is then handed to the MPDATA routine as specific humidity. With regard to Eq. 6.2, this effectively means that the mass (density) information of the advected variable is partly incorporated in G, partly in $\mathbf{v}_a^{n+1/2}$. Further options of the MPDATA routine were specified to gain a second-order accurate sign and monotonicity-preserving advection scheme (IORD0=2,ISOR=2,NONOS=1,IDIV=0, see also Smolarkiewicz 2005).

Idealised simulations were carried out with the CHRM model to test the correct implementation (D. Lüthi, *pers. comm.* 2005). Blattmann-Singh (2005) compared an STW tracer implementation using different advection algorithms (upstream, leap frog, Lax-Wendroff, MPDATA) and initialisations in deformative and translational 2D flow fields. Further experiments were performed to test the agreement between the original 3D leap-frog advection and MPDATA tracer advection in a setup with a rotational wind field, and in uniform flow over an isolated hill (not shown).

Cloud water advection

A particular problem arises from the advection of cloud water. For a field with smooth gradients, the leap-frog advection produces acceptable solutions. However, as the 1D advection experiments in Chapter 5 and the 2D-advection study by Blattmann-Singh (2005) have shown, considerable numerical noise is generated from leap-frog advection if the advected quantity exhibits strong gradients on small spatial scales ($\sim 2 - 5\Delta x$). While the water vapour field has mostly small gradients, the cloud water field is very spotty, and has large areas where no liquid water is present. This is quite similar to the problems one would have when using the leap-frog advection for a tagging implementation. Hence, the potential mismatch between the background and the tracer field for cloud water is rather large when using different advection schemes, and losses of tagged cloud water due to overshoots become more likely (see Sec. 5.1.3).

In the future, it may prove favourable to improve the tagged cloud water advection, either by treating cloud water as a parallel rather than a synchronised tracer (which would however raise problems for the precipitation parametrisations). Another option is to use MPDATA for both the background and the tagged cloud water field, which would increase numerical consistency and tremendously reduce the potential for overshoots.

Background moisture field advection with MPDATA

MPDATA was in addition implemented for the advection of the background moisture field in the CHRM. This was motivated by two reasons:

(i) Testing for a correct implementation of the advection algorithm is difficult in a STW implementation, as any errors during implementation may be hidden by the synchronisation with the background field after each time step. Using MPDATA for both the tracer and the background moisture, numerical consistency is achieved, which during the implementation allows an immediate attribution of errors to other causes, such as parametrisations.

(ii) The advantageous properties of the MPDATA advection scheme (conservation of mass, preservation of sign, less numerical noise) should also offer benefits for the overall performance of the CHRM. This prospect is however moderated by the likely requirement of a new model tuning, in particular with respect to the precipitation parametrisations.

Changes in the code related to the implementation of this feature were mostly carried out in the routine `progexp`. In addition, Asselin filtering had to be deactivated (routine `asselin`).

A 72 h model simulation with the MPDATA advection scheme (namelist parameter *qalg*=1) produces results which are quite similar to the standard leap-frog advection scheme (*qalg*=0) (Fig. 6.3). However, more structure is visible in the moisture field with MPDATA advection. In addition, dry/wet sectors partly have differing magnitudes, and precipitation is more widespread in the MPDATA simulation. This shows that (a) the 3D advection and horizontal/vertical diffusion are correctly implemented, (b) that the additional fine-scale structure of the moisture field feeds back into the dynamics of the model and creates different simulation results. Further model validation would have to be carried out to clarify if this hybrid CHRM-MPDATA model can actually improve the model simulations.

6.2.2 Horizontal diffusion

Diffusion is as much a parametrisation for subgrid-scale eddy mixing as a numerical cure for the weaknesses of the advection scheme. Hecht et al. (1995) noted that the leap frog algorithm is unsuitable for pure advection problems, and only applicable to combined advection and diffusion problems. Explicit diffusion selectively dampens small-scale numerical ripples created by the leap-frog advection scheme. The CHRM applies a 4th-order linear diffusion in the horizontal (D_4). Near the boundaries, due to the shorter fetch, a 2nd-order linear diffusion (D_2) is applied:

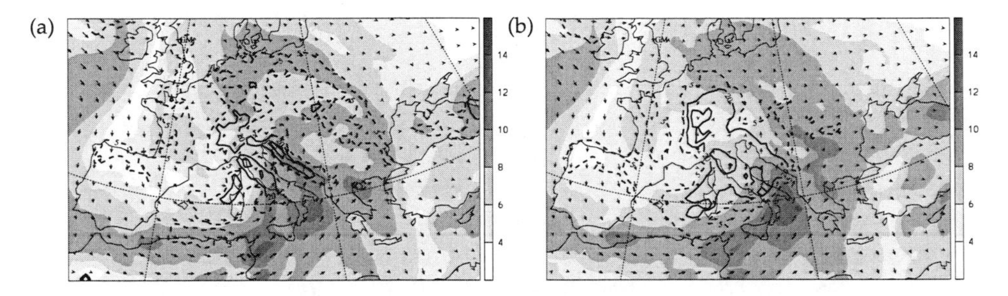

Figure 6.3: Specific humidity at 700 hPa (shaded) and accumulated total precipitation (contours) after 72 h of model integration time for the Elbe flood case (00 UTC 13 August 2002, see Chapter 7) with (a) CHRM leap-frog advection, (b) MPDATA moisture advection.

$$\begin{cases} D_4 = -K_4 \nabla^2 (\nabla^2 q_v), \\ D_2 = K_2 \nabla^2 q_v. \end{cases} \tag{6.5}$$

In order to avoid unrealistic negative values near cloud edges, only q_v is diffused horizontally. The diffusion coefficients K_4 and K_2 are set according to the equations

$$\begin{cases} 0 \leq K_4 \leq \frac{1}{128 \Delta t}, \\ 0 \leq K_2 \leq \frac{1}{16 \Delta t}. \end{cases} \tag{6.6}$$

For use in the CHRM, due to the preferred damping at short wavelengths, the values

$$\begin{cases} K_4 = \frac{1}{2\pi^4 \Delta t} \\ K_2 = \frac{1}{2\pi^2 \Delta t} \end{cases} \tag{6.7}$$

are recommended (DWD 1995). The horizontal diffusion coefficients of moisture are set to 20% of the momentum diffusion (Eq. 6.7).

As suggested by the 1D tagging experiments (Sec. 5.1.5), horizontal diffusion in combination with MPDATA will lead to the generation of overshoots in the tracer field and hence to the loss of tracer mass. While there is an option to diffuse the tagged water vapour field with a horizontal diffusion coefficient set to 10% of the momentum diffusion, this option is currently not used. However, as MPDATA produces less implicit diffusion than for example the leap-frog advection scheme, in the future it might prove physically more consistent to apply an explicit diffusion to the tagged water vapour field. Currently, the tagged tendency due to horizontal diffusion is resolved during the implicit vertical diffusion (see below). It is however also possible to add the tendency to the moisture field directly when calling the MPDATA routine (P. Smolarkiewicz, *pers. comm.* 2005). As for the background field q_c, tagged cloud water is not explicitly diffused horizontally.

6.2.3 Vertical diffusion

Vertical diffusion is a crucial step for the transport of (tagged) water vapour after evaporation from the surface to higher levels of the model atmosphere. Vertical advection is a rather inefficient transport process at low model levels, and without the vertical diffusion artificially strong gradients would build up near the surface. In a simulation of the Elbe flood case (see Chapter 7) where the vertical advection and diffusion of temperature and moisture had been switched off, a pronounced upper-level PV streamer to the west of the Alps died away in a matter of hours (not shown). Hence, vertical diffusion is a very important component of the model's water and energy cycle, and the choice of the turbulence parametrisation can have a tremendous impact on the simulation.

In the free atmosphere, vertical eddy diffusivity K_H^V is parametrised by a 2nd-order turbulence kinetic energy (TKE) closure (Mellor and Yamada 1974; DWD 1995). In the Prandtl layer, vertical turbulent transport is parametrised with a transfer coefficient C_H^V according to Louis (1979); Louis et al. (1982). The corresponding diffusion equations are:

$$g\left(\frac{\partial q}{\partial \eta}\right)^{-1}\frac{\partial q}{\partial \eta} = \begin{cases} g\rho^2 \cdot K_H^V \cdot \frac{\partial q_v}{\partial p} & \text{for the free atmosphere, } k < KE \\ \rho_s \cdot C_H^V \cdot \frac{\partial q_v}{\partial p} & \text{for the boundary layer, } k = KE. \end{cases} \tag{6.8}$$

where ρ_s is the density of air at the surface, k is the model level index, and KE is the lowest model layer.

For stability reasons, vertical diffusion and advection are calculated implicitly in the CHRM numerics. The implicit formulation results in a system of algebraic equations which is solved by Gaussian elimination (e.g. Schär 2002). The implicit formulation also redistributes the tendencies due to horizontal diffusion (q_v), grid-scale precipitation (q_v and q_c), and convective precipitation (q_v).

In order to implement WVTs, the implicit code has to be stripped from the vertical advection terms so that only the vertical diffusion terms are retained. In addition, the code has to be changed from a leap-frog time step to a forward time step. The remainder is an implicit solution for the vertical diffusion, which also incorporates the tagged tendencies from horizontal diffusion, grid-scale, and convective precipitation. The major advantage of this tagging implementation is that the same diffusion coefficients (but for time level n instead of $n-1$) can be used for the WVT, and numerical consistency is guaranteed. A possible disadvantage is that in order to obtain smooth results, the vertical diffusion coefficients were chosen unphysically large, and hence the vertical diffusion of tracer could be exaggerated.

Technically, the code for vertical diffusion is split into the determination of the transfer coefficients for atmospheric and surface turbulence (`partura` and `parturs`) and the actual application as part of the routine `progexp` (Table 6.1). For the tagging implementation, the transfer coefficients remain unchanged, and all code changes are carried out in the routine `progexp`.

During the implicit vertical advection/diffusion of the background moisture fields, occasionally negative values are generated which are filled by moisture 'borrowed' from the neighbouring grid cells below. The moisture transfer due to this 'gap-filling' could in principle also be imitated by the tagged moisture field. However, since it is without physical meaning, such a procedure is questionable for the tracer field. Hence, overshoots generated by this process are accounted as lost water mass. Similarly, during the vertical diffusion of tagged water vapour, occasional negative values are generated, which are then set to zero and accounted as extra mass.

A possible alternative treatment of the vertical diffusion is offered by using a modified vertical velocity $\tilde{\eta}$ for the MPDATA advection (P. Smolarkiewicz, *pers. comm.* 2005):

$$\tilde{\eta} = \dot{\eta} - \frac{K_H^V \cdot \nabla q}{q} \tag{6.9}$$

In this case, it would also be required to incorporate the tendencies due to other physical parametrisations directly during the MPDATA advection. Given that the same diffusion coefficients can be used, this could in the future offer an elegant alternative to the implicit algorithm with its particular shortcomings.

6.2.4 Lateral boundary relaxation

The boundary relaxation following Davies (1976) imposes a gradually decreasing forcing from the boundaries to the inner domain. The boundary fields are linearly interpolated between two subsequent (typically 6 h) initialisation fields, and 'relaxed' (i.e. communicated into the model domain) with decreasing weight further from the boundaries according to the relaxation coefficient μ (Fig. 6.4).

The relaxation function quickly decreases away from the boundaries. Typically, the 8 grid points at the boundary are considered as adaption zone where the model does not contain fully consistent meteorological information. Hence, for the tagging implementation, the boundary region is excluded from all budget calculations (see Sec. 6.1.3). It should be noted, however, that the actual relaxation zone covers 19 grid points at each boundary, even though values of μ are small (Fig. 6.4b).

The actual relaxation is carried out in the routine `lb_relax` (see Table 6.1). The implementation of the tracer field follows the same concept as for the other relaxed fields. Hence, the relaxation equation for the tracer field at each grid point i, j, k is

$$f_{i,j,k}^r = f_{i,j,k} - \mu_{i,j} \cdot (f_{i,j,k} - f_{i,j,k}^b) \tag{6.10}$$

where μ is the relaxation coefficient, f is the fraction of marked water, the superscript r indicates the relaxed value, and the superscript b the boundary relaxation value. The relaxation of the tracer field differs from the relaxation of other fields insofar as the tagged fraction is relaxed rather than the actual quantity. The boundary field f^b is

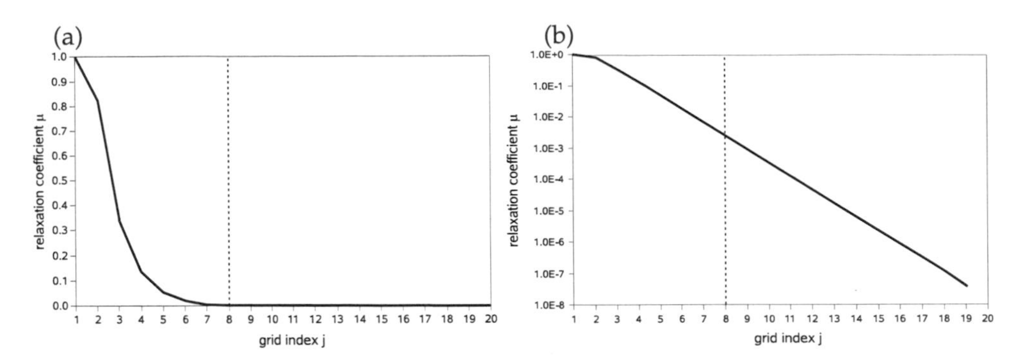

Figure 6.4: Function of the lateral boundary relaxation coefficient μ vs. horizontal grid index j; (a) normal, (b) logarithmic y-axis scaling. The usual boundary cut-off at $j = 8$ is indicated by a dashed line.

defined by the tracer initialisation field described in the input file (see Sec. 6.1.2), but without linear interpolation in time. This is effectively the same as relaxing the tagged water field against a corresponding water vapour density. In principle, this allows for future nesting applications with WVT in the CHRM.

As mentioned above, the relaxation zone, at least numerically, extends 19 grid points into the domain. Even though values are small, it is possible that numerical noise is introduced in the model domain, and may even be reflected at the boundaries. The tracer field, which at least initially may contain areas where no tracer is present, is particularly vulnerable to such numerical effects. A possible modification of the relaxation code to reduce this risk would be to set the relaxation coefficient to exactly zero outside the 8 grid points wide relaxation zone.

6.3 Phase change parametrisations

At the temporal and spatial scale of the CHRM, physical processes which involve phase changes can not be fully resolved, and therefore have to be covered by numerical parametrisations. Typically, in a RCM, these parametrisations can become considerably more complex than in GCMs. This includes condensation of water vapour and evaporation of cloud water, grid-scale precipitation, convective precipitation, and soil processes.

Two particular problems occur for the implementation of a tagging procedure into the physics parametrisations: (i) Specific thresholds may exist, which must be exceeded in order to trigger a process. If simply the identical parametrisation routine was applied to the tagged moisture field, condensation for example could occur in the main moisture field, but not necessarily in the tagged moisture field. (ii) Specific processes may feed back into the dynamics by latent heat release. Running identical parametrisation routines for the tagged as for the main moisture field might produce additional latent heat release and hence spurious temperature perturbations.

The solution to these problems for the WVT implementation lies in mimicking the water mass transfer of the main moisture field. All tagged moisture transfer due to phase changes are considered as proportional to the fraction of tracer in the source reservoir. This allows for a similar treatment of all phase change parametrisations. The problem reduces to extracting the mass fluxes from the parametrisations. Technically, for each mass transfer $\Delta\rho_x$ from reservoir x which is due to phase changes, the tagged mass flux $\Delta\rho_{x,t}$ is assumed to be proportional to $\Delta\rho_x$ and the fraction of tagged mass in reservoir x:

$$\Delta\rho_{x,t} = \Delta\rho_x \frac{\rho_{x,t}}{\rho_x}. \tag{6.11}$$

This general principle is in the following applied to implement the tagging procedure into the parametrisations of cloud water formation (condensation), grid-scale precipitation, convective precipitation, and soil moisture.

6.3.1 Condensation and evaporation

Condensation of q_v and evaporation of q_c is the grid-scale mechanism for mass exchange between the atmospheric water phases in the CHRM. The condensation parametrisation requires that water vapour and cloud water be in thermodynamic equilibrium after each time step. Hence, temperature and moisture (i.e. the respective saturation vapour pressure at each grid cell) determine the fraction of water vapour which is converted to cloud water. In supersaturated conditions (defined here always as RH $> 100\%$ with respect to water), all the excess water vapour in a grid cell is immediately transferred to the liquid phase. As there is no explicit ice phase, ice supersaturation is non-existent in the CHRM microphysics. Technically, the iterative algorithm splits the total water and enthalpy in a grid cell according to the imposed saturation limit into the components T, q_v and q_c. There is no cloud ice phase in the CHRM on the grid scale (in contrast to the grid-scale precipitation parametrisation, see below).

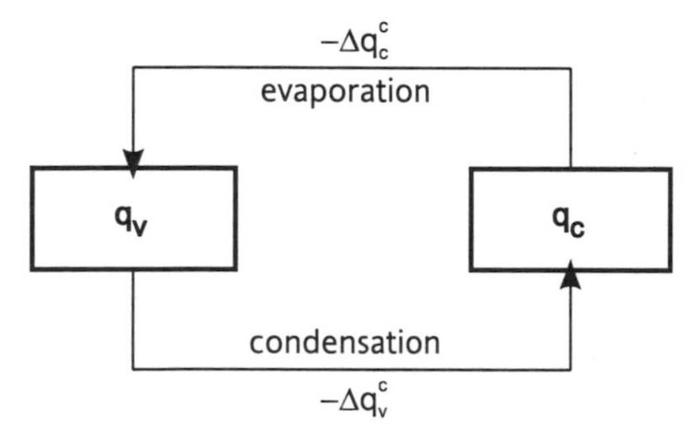

Figure 6.5: Condensational and evaporative mass transfer between water vapour q_v and cloud water q_c in the condensation parametrisation. See text for details.

The tagging implementation directly mimicks the condensational mass exchange of total water. By comparing q_v and q_c in each grid cell before and after condensation, the mass transfer to both phases resulting from condensation or evaporation is calculated as

$$\begin{cases} \Delta q_v^c = (q_v^c - q_v), \\ \Delta q_c^c = (q_c^c - q_c). \end{cases} \tag{6.12}$$

where x^c denotes the quantity x after condensation or evaporation (see Fig. 6.5). For consistency, it is checked whether both fluxes add up to zero. According to Eq. (6.11), the tagged moisture tendencies $\Delta q_{v,t}$, $\Delta q_{c,t}$ are then calculated as

$$\begin{cases} \Delta q_{v,t} = \Delta q_v^c \frac{q_{v,t}}{q_v}, \\ \Delta q_{c,t} = \Delta q_c^c \frac{q_{c,t}}{q_c}. \end{cases} \tag{6.13}$$

Finally, the tagged specific humidity is converted to a water vapour density $\Delta\rho_{x,t}$. This tagged moisture tendency is added to the tagged water vapour field $\rho_{v,t}$ and subtracted from the tagged cloud water field $\rho_{c,t}$ after the horizontal and vertical advection and diffusion of those fields.

Hence, following the advection and diffusion of tagged water vapour, the tagging condensation routine carries out the following steps for each grid cell:

1. Save the pre-condensation values of q_v, q_c, T
2. Do condensation as in original CHRM code
3. Calculate moisture tendency due to condensation from Eq. 6.12
4. Consistency check if $\Delta q_v^c = -\Delta q_c^c$
5. Depending on evaporation ($\Delta q_v^c > 0$) or condensation ($\Delta q_c^c > 0$) calculate the tagged tendency from Eq. 6.13

Condensational mass exchange is calculated at three instances during each time step: (i) after vertical advection, (ii) after the semi-implicit gravity-wave corrections, and (iii) after the Asselin filtering and the lateral boundary relaxation[1] (see Table 6.1).

[1]In case (iii), only fields from time levels n and $n-1$ are affected, which is ignored for the WVT. In case (i) the condensation routines were originally calculated level by level jointly with the 'gap filling' algorithm in the routine `progexp`. The inline condensation code was replaced by a call to the condensation routine. This change to the original code has however no consequence for the untagged model output.

6.3.2 Grid-scale precipitation

The parametrisation of grid-scale precipitation calculates the tendencies of T, q_v, q_c and the surface rain and snow rates R_r, R_s which are due to large-scale precipitation processes. The implementation in the CHRM model is a Kessler (1969)-type parametrisation. While there is no treatment of cloud ice in the advection and condensation processes, a simple distinction between precipitation from warm and cold clouds (but excluding mixed clouds) is used. The parametrisation redistributes water mass between the four reservoirs q_v, q_c, R_r and R_s as shown in Fig. 6.6. The related fluxes between reservoirs are denoted by $\mathcal{F}_{xy}$, where x and y denote the source and target reservoirs (v =water vapour, c =cloud water, r =rain, s =snow). Note that q_c acts as source reservoir only, and that vapour can contribute to the snow reservoir also in warm clouds, as indicated by the dotted lines.

Depending on the temperature and cloud water conditions, different cloud microphysical processes, represented by the fluxes $\mathcal{F}_{xy}$, are active. One general distinction is made between cold conditions ($T < T_{\text{melt}}$) and warm conditions ($T \geq T_{\text{melt}}$). A second distinction is made between cases where clouds are present ($q_c > 0$) and such where no clouds are present, but rain from above is falling through a grid cell ($q_c = 0$ and $R_s + R_r > 0$). Hence, the CHRM parametrisation distinguishes between 4 situations, which are each characterised by specific processes and the corresponding fluxes between reservoirs (Fig. 6.6):

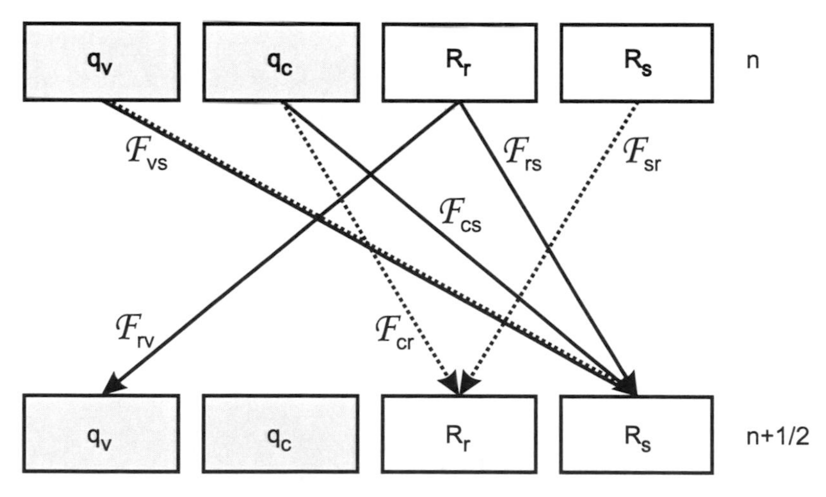

Figure 6.6: Fluxes between water reservoirs in the grid-scale precipitation parametrisation. The grid-scale precipitation parametrisation redistributes moisture at time level n (top row) to time level $n + 1/2$ (bottom row). Cold processes are denoted by solid and warm processes by dotted lines. See text for details.

1. Cold clouds ($T < T_{\text{melt}}$ and $q_c > 0$)

   $\mathcal{F}_{rv}$ = evaporation
   $\mathcal{F}_{vs}$ = deposition
   $\mathcal{F}_{cr}$ = autoconversion + accrescence
   $\mathcal{F}_{cs}$ = riming + nucleation
   $\mathcal{F}_{rs}$ = freezing rain

2. Warm clouds ($T \geq T_{\text{melt}}$ and $q_c > 0$)

   $\mathcal{F}_{cr}$ = autoconversion + accrescence + shedding
   $\mathcal{F}_{sr}$ = melting snow

3. No clouds, cold precipitation ($T < T_{\text{melt}}$ and $q_c = 0$ and $R_s + R_r > 0$)

   $\mathcal{F}_{vs}$ = deposition
   $\mathcal{F}_{rv}$ = evaporation
   $\mathcal{F}_{rs}$ = freezing rain

4. No clouds, warm precipitation ($T \geq T_{\text{melt}}$ and $q_c = 0$ and $R_s + R_r > 0$)

   $\mathcal{F}_{vs}$ = deposition
   $\mathcal{F}_{rv}$ = evaporation
   $\mathcal{F}_{sr}$ = melting snow

Again, rather than carrying out the precipitation parametrisation on the tagged moisture field, the tagging implementation mimics the mass transfer from the main moisture field. For each total flux $\mathcal{F}_{xy}$ that occurs in a specific situation, the respective tagged flux $\mathcal{F}_{xy,t}$ is calculated according to:

$$\mathcal{F}_{xy,t} = \mathcal{F}_{xy} \frac{q_{x,t}}{q_x}. \tag{6.14}$$

The algorithm for calculating the grid-scale precipitation proceeds layer by layer from top to bottom, thereby also integrating the precipitation rates from top to bottom. At the end of each vertical loop, the tagged moisture tendencies are converted to water vapour density.

The routine `pargsp` experienced a number of changes in the code to accommodate tagged water. These include the additional input fields $q_{v,t}$, $q_{c,t}$ and the new output fields $\Delta q_{v,t}^{\text{gsp}}$, $\Delta q_{c,t}^{\text{gsp}}$ for the tagged moisture tendencies and $R_{r,t}^{\text{gsp}}$ and $R_{s,t}^{\text{gsp}}$ for the precipitation rate of tagged rain and snow due to grid scale processes. The tagged moisture tendencies are added to the tagged moisture fields during the vertical diffusion, analogously to the proceeding for the main moisture fields.

As described in Section 6.1.4, the corresponding GRIB variables DQV_WGSP, DQC_WGSP, PRR_WGSP and PRS_WGSP are available as output fields (Table 6.3). Finally, for performance increase, the calling sequence of `pargsp` was changed, so that the complete layer is treated by one call, instead of the previous line-wise calling to save local storage. This is however a mere matter of performance, and does not affect the grid-scale precipitation itself.

6.3.3 Convective precipitation

Convection is of crucial importance for the vertical redistribution of heat and moisture in the troposphere. Typical effects of convection are a drying of the lower troposphere, heating the middle troposphere, and upward transport of momentum, in particular in the tropics. At sufficiently fine grid-scales (< 3 km), convection is more or less explicitly resolved in numerical models. At the typical grid scales of most current NWP and RCM models however, this sub-grid scale process has to be parametrised.

The CHRM convection scheme

The parametrisation of convective precipitation in NWP models is a complex task, as regionally, specific types of convection can occur, that are governed by differing cloud dynamical processes. The CHRM model uses an implementation of the one-dimensional mass flux convection parametrisation proposed by Tiedtke (1989), which is also currently employed in the ECMWF model. The Tiedtke (1989) scheme is a bulk scheme (no cloud spectra) and acts on one atmospheric column at a time. It has been designed to represent three different types of convection: (i) deep (penetrating) convection, typically occurring in the tropics, (ii) shallow cumulus convection, which for instance leads to tradewind cumuli, and (iii) mid-level convection, which is typically associated with mid-latitude frontal systems and systems of organised convection. For an overview of the dynamics of convective clouds see e.g. Houze Jr. (1993).

All three types of convection include representations of updraughts and downdraughts, entrainment and detrainment of environmental air, cloud water, and the evaporation of rain (Fig. 6.7). The driving mechanism (closure assumption) of the scheme is in all cases the influx of mass at the cloud base. This mass influx then triggers ascent, and finally descent to the level of free sinking (LFS). The closure assumption in the cases (i) and (iii) is that the mass flux is caused by low-level moisture convergence of the large-scale flow (and uplift), while for (ii) the mass flux is equal to the turbulent moisture influx from the PBL.

The implementation of the parametrisation is organised in the subroutine `parcon` (see Table 6.1). Calling several subroutines, the parametrisation proceeds along the following steps (Tiedtke 1989):

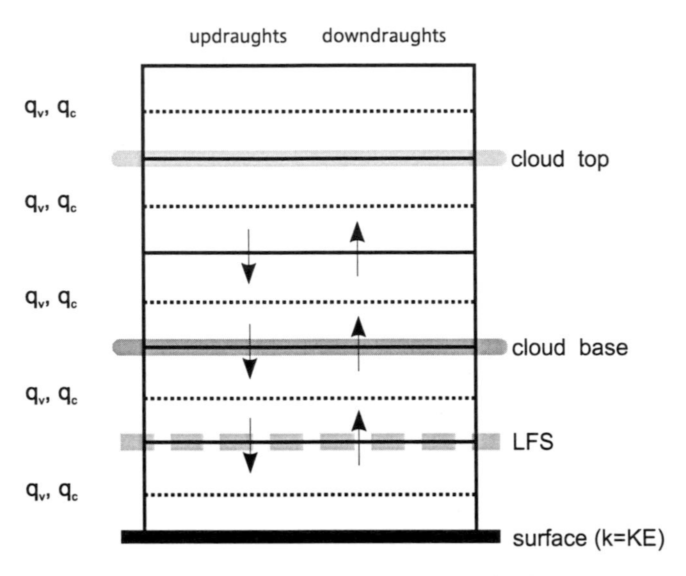

Figure 6.7: Sketch of a grid column with parameters relevant for the convection parametrisation scheme. LFS = level of free sinking. See text for details.

1. Definition of constants and parameters in `cu_const`
2. Initialisation of half level values of, among others, T, q_v, u, v and of updraught and downdraught values in `cu_ini`
3. Determination of cloud base (for deep and shallow convection) in `cu_base` and of cloud base mass flux from the moisture budget in the PBL
4. Cloud ascent calculations in `cu_asc` (first call) in absence of downdraughts
5. Downdraught calculations:
   - values of T, q_v, u, v at LFS in `cu_dlfs`
   - determination of moist descent in `cu_ddraf`
   - recalculation of cloud base mass flux including the effect of cumulus downdraughts
6. Final cloud ascent calculation in `cu_asc` (second call) and determination of cloud top
7. Final adjustments to convective fluxes in `cu_flux` and evaporation in sub-cloud layer
8. Calculation of surface rain/snow rates R_r^{con} and R_s^{con} and the tendencies of T and q_v in `cu_dtdq`
9. Calculation of the tendencies of u and v in `cu_dudv`

The moisture tendency due to convection Δq_v^{con} is then fed back into the large-scale moisture field during the vertical diffusion (see Sec. 6.2.3). Note that there is no large-scale cloud water tendency due to convection.

Tagging implementation of the convection scheme

A straightforward application of the convection scheme to the tracer field would lead to inconsistent triggering between tracer and background moisture. Hence, the thresholds and calculations of cloud properties in the parametrisation render such a direct application impossible. The Tiedtke (1989) scheme is however fully mass conservative, and therefore again offers the possibility to mimic the vertical redistribution of water vapour by the tagged moisture field. There are two options for implementing the tracer advection in the parametrisation scheme:

(i) Bulk method. Under the assumption that all moisture is well-mixed within a convective cloud, there is no need to separate the contribution of tracer water from each layer. The method can hence proceed as follows:

1. Determine the cloud base and top

2. Calculate an average tagged moisture fraction for the whole cloud column as

$$f^{\text{con}} = \frac{\sum_{\text{base}}^{\text{top}} (q_{v,t} + q_{c,t})}{\sum_{\text{base}}^{\text{top}} (q_v + q_c)} \tag{6.15}$$

3. Calculate mass fluxes and precipitation from the convection scheme

4. Calculate the tagged moisture tendencies $\Delta q_{v,t}^{\text{con}}$ from

$$\Delta q_{v,t}^{\text{con}} = f^{\text{con}} \cdot \Delta q_v^{\text{con}} \tag{6.16}$$

5. Calculate the tagged precipitation rates R_r^{con} and R_s^{con} from

$$\begin{cases} R_{r,t}^{\text{con}} = f^{\text{con}} \cdot R_r^{\text{con}} \\ R_{s,t}^{\text{con}} = f^{\text{con}} \cdot R_s^{\text{con}} \end{cases} \tag{6.17}$$

(ii) Flux method: In a more detailed approach one could use the final values for updraughts and downdraughts of moisture calculated by the parametrisation to calculate mass fluxes for tagged water according to Eq. (6.11). A tedious task is however to mimick all fluxes from and to the precipitation throughout the convection parametrisation. It is not clear to what extent water transport due to convection will be more realisticly captured by the flux method.

In this tagging implementation, the simpler bulk method (approach (i)) is selected. Future work could evaluate the effects of using a more detailed treatment of convection with the flux method.

6.3.4 Soil model

The CHRM has an advanced soil model based on the work of Dickinson (1984) and Jacobsen and Heise (1982). In climate simulations, this soil model is capable of retaining realistic soil moistures over several consecutive years (Vidale et al. 2002). Even though the soil plays an important role in the atmospheric water cycle, for example in acting as an intermediate storage or in determining surface evaporation fluxes, the soil model is not treated in detail at this stage of tagging implementation in the CHRM.

Currently, the only contribution from the soil is that during slab initialisation the tagged moisture directly above the surface is used as a lower boundary condition in the vertical diffusion scheme (see Sec. 6.2.3). Over sea, the moisture directly above the surface is always assumed to be saturated, while over land, this surface moisture is a function of soil type, soil water, and plant cover. No tagging information is retained from precipitation that infiltrates into the soil storage or that contributes to runoff.

In the future, implementing water vapour tagging in the soil model could be an important contribution to closing the tagged water cycle in CHRM. Studies could then for example consider re-evaporating tagged precipitation (possibly with a different tracer identifier), assess the moisture recycling in an area, keep track of marked water in the compartments of the soil model, and evaluate the contribution of tagged water to runoff.

Building on this detailed description of the implementation of WVT in the CHRM, the next chapter provides an illustrative application of the CHRM-WVT. The extreme floods in central Europe during August 2002 (Elbe flood) are considered with the aim (i) to provide an example for WVT in the CHRM as described in this chapter, (ii) to investigate the reliability of the method, and (iii) to study the moisture sources which contributed to this extreme event.

Chapter 7

Tagging Study of the August 2002 Flood in Central Europe

The August 2002 flood of the Elbe river was one of the worst natural catastrophes in Europe in recent decades. The flood affected a large part of central Europe, but most severely it struck southern and eastern Germany, Austria, and the Czech Republic. After long and intense rainfalls with a maximum over the headwaters of the Elbe river, the city centres of Dresden and Prague were flooded. With 312 mm of precipitation at Zinnwald-Georgenfeld in the Erzgebirge on 12 August 2002, the torrential rains set a new 24 h accumulated precipitation record for Germany. Damage estimates ranging between 12-20 billion Euros place the August 2002 flood (Elbe flood) first on the list of the most costly natural disasters in Europe, taking over from the winter storm *Lothar* in 1999 (Marsh and Bradford 2003; Ulbrich et al. 2003a).

The flash floods started on 12 August, after two intensely precipitating cyclones had travelled first along the southern fringe of the Alps and then northward over Austria on 6/7 August and 11/12 August, respectively (James et al. 2004; Ulbrich et al. 2003a). On 10 August, the large-scale flow over Europe featured a large quasi-stationary stratospheric PV streamer upstream of the Alps, that led to (i) low-level cyclogenesis in the lee of the Alps (Bleck and Mattocks 1984; Steinacker 1984; Tafferner 1990; Appenzeller et al. 1996; Morgenstern and Davies 1999) and (ii) the advection of moist warm air from the Eastern Mediterranean into central Europe. The path of these cyclones is sometimes classified as a 'Vb-track' (van Bebber 1891), a track that is known to be associated with heavy precipitation in central Europe (Ulbrich et al. 2003b). Better understanding of the reasons for the flood does however not ensue directly from such a classification.

Investigations of the flood origin focused primarily on precipitation mechanisms and additional enhancing factors. Ulbrich et al. (2003a) noted the strong impact of frontal uplift at a slow-moving convergence line. While orographic enhancement most certainly helped setting the high precipitation records at the Erzgebirge, the role of convection is less clear (Zängl 2004). James et al. (2004) hypothesised that above-average sea surface temperatures (SST) in the eastern Mediterranean and Black Sea enhanced evaporation. Furthermore, it has been hypothesised that the filling of the soil water storage due the earlier cyclone was important for the later widespread flooding (Marsh and Bradford 2003; Stohl and James 2004; James et al. 2004). Recycling of moisture seemed however less important, as Zängl (2004) concluded from the small sensitivity of simulated precipitation magnitudes to the soil moisture initialisation.

We chose the Elbe flood period (00Z 10 August to 00Z 13 August 2002) as a first application of the CHRM tagging methodology for several reasons. First, the event is well-represented in the literature. In addition to detailed synoptic accounts (Ulbrich et al. 2003a,b; James et al. 2004), a sensitivity study of a regional NWP model simulation of the event (Zängl 2004) and an investigation of the moisture sources with a Lagrangian methodology (James et al. 2004; Stohl and James 2004) are available. Despite these research efforts, open questions remain with respect to the moisture sources for the event. While James et al. (2004) and Stohl and James (2004) found contributions from numerous sources to the precipitation event, no quantitative and unequivocal attribution of moisture sources was achieved. Furthermore, since convective precipitation seems to have played a role, it limits the applicability of most Lagrangian diagnostics for that specific event. Finally, it remains to be resolved what the individual contributions of enhanced SST, land evaporation, large-scale moisture transport, and the various large water basins (Atlantic Ocean, Mediterranean, Black Sea) to the flood were. The tagging methodology should be able to reveal more insight into the processes that gave rise to the devastating floods in August 2002.

7.1 CHRM simulation of the event

The CHRM model simulation was set up for a domain covering central and southern Europe, and including the whole Mediterranean and a part of the North Atlantic ocean (Fig. 7.1 and Table 7.1). For this initial study, a relatively coarse horizontal resolution of ~55 km was chosen. The orography at this resolution is necessarily smoothed to some degree, but highest peaks still reach ~2500 m in the Alps, the Atlas mountains, and the Caucasus. The Erzgebirge, where the largest precipitation records were observed during the Elbe flood, appears as a broad area with altitudes of about 400–600 m altitude in Fig. 7.1. An enhanced vertical resolution (28 instead of 20 levels) was chosen, compared to the standard CHRM setup, to better resolve low-level moisture transport. Apart from the 30% reduced time step, other options in Table 7.1 are comparable to those chosen in other CHRM/HRM simulations (e.g. Lüthi et al. 1996; Vidale et al. 2003).

The simulation of the case is first compared with the ECMWF analyses which were used to force the RCM at the boundaries. Fig. 7.2 displays the large-scale circulation in terms of potential vorticity (PV) at 325 K together with the equivalent potential temperature (θ_e) at 850 hPa. A stratospheric PV streamer visible upstream of the Alps on 00Z 10 Aug sheds a cut-off 24 h later. The cut-off moves cyclonically around the Alps and first elongates (00Z 12 Aug) and then dissolves while spiralling into central Eu-

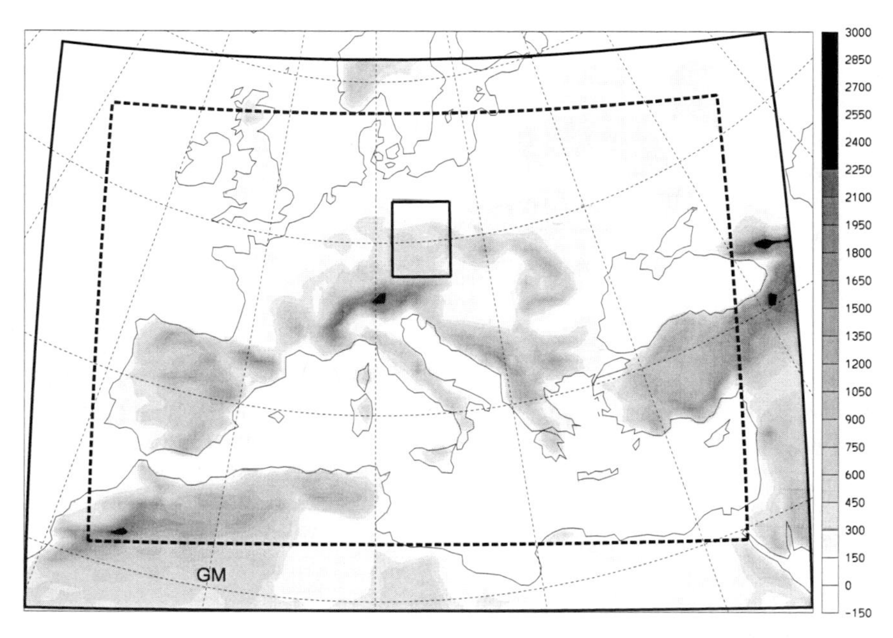

Figure 7.1: CHRM simulation domain with inner domain where boundary relaxation influences are negligible (dashed), and precipitation target area over the Elbe headwaters (~11.5–17.3°E; 48–52.3°N;)

Table 7.1: Setup of the CHRM model for the simulation of the August 2002 floods.

Parameter	Settings
Simulation domain	27°W 55°N; 55°E 55°N; 33°E 32°N; 10°W 32°N
Time period	00 UTC 10 Aug – 00 UTC 13 Aug 2002
Time step (s)	200
Integration time (h)	72
Domain size (grid points)	91x65x28
Horizontal resolution (°)	0.5 (~55 km)
Vertical layer positions (hPa)	995, 975, 950, 925, 900, 875, 845, 815, 785, 750, 715, 680, 640, 600, 550, 500, 460, 420, 380, 340, 300, 270, 240, 210, 180, 140, 100, 60
Boundary update interval (h)	6
Moisture advection	leap-frog scheme
Parametrisations and settings	surface-layer turbulence, atmospheric turbulence, convective precipitation, grid-scale precipitation, radiation parametrisation, Dickinson-type soil model, radiative upper boundary, semi-implicit corrections

rope (00Z 13 Aug). While the position of the cut-off agrees well between analysis and simulation, differences in size and shape are apparent. In particular, the positive PV anomaly is larger throughout the CHRM simulation. The response of the lower troposphere to the upper-level circulation is a southerly advection of moist/warm air into central Europe to the east of the PV anomaly along a cyclonically curved path. There is a good overall correspondence in the position of the low-level fields between analysis and simulation. A slight tendency of the simulated θ_e field towards higher maximum values and a further northward progression of high θ_e air can however be noted. In summary, the period shows a high correspondence with the analysis field. Therefore, the CHRM simulation provides a good representation of the real atmosphere, and a viable basis for a first RCM tagging analysis of the atmospheric water cycle.

A second validation compares the simulation directly with observed (rather than assimilated) data. Moisture composition in the upper troposphere observed from Meteosat satellites (wavelength 5.7–7.1 μm) is compared with the simulated water vapour (WV) in a section of the upper troposphere (Fig. 7.3, compare also Appenzeller and Davies (1992)). The sequence shows a significant change from mostly wet to mostly dry conditions during the study period. On 00Z 10 Aug, there is already a large moisture plume visible above the Balearic islands (green spot), indicating heavy precipitation in that area (James et al. 2004). The simulation shows this plume as the spearhead of a large moisture tongue reaching from the Atlantic towards Spain. Reflecting further convective activity over the Mediterranean, the plume both in the satellite image and the simulation gains further moisture while spiralling into eastern Europe during 11 Aug. At the end of the study period on 00Z 13 Aug, a clear boundary separates dry from moist air masses across Germany in the WV image. In the simulation, the southern extent of this boundary is displaced ~100 km to the north, and a moisture plume has appeared near the Atlas mountains which is not observed from satellite. Apart from these differences, the simulated and observed WV fields agree closely. However, the upper troposphere is not the dominant moisture source (Newell et al. 1992), and reflects only some of the moist processes in the lower troposphere (in particular convection).

Next, the precipitation of the CHRM simulation is compared with data from the literature. Fig. 7.4a shows the simulated precipitation field, accumulated over the last 48 h of the simulation (00Z 11 Aug – 00Z 13 Aug). This period coincides with the strongest precipitation intensities in the Erzgebirge (Ulbrich et al. 2003a). Peak simulated precipitation exceeds 150 mm near 14.3°E; 49.9°N and 80 mm in a north-south extended area in the target region (Fig. 7.1). The location of the precipitation maximum hence corresponds reasonably well with observations (Zängl 2004, his Fig. 3d). Precipitation maxima are clearly lower than observed, which is not surprising considering the relatively coarse resolution of the simulation.

A separation into contributions from simulated large-scale (Fig. 7.4b) and convective precipitation (Fig. 7.4c) shows that the precipitation maximum in the Erzgebirge as well as at the Alpine range originates for the most part from large-scale condensation.

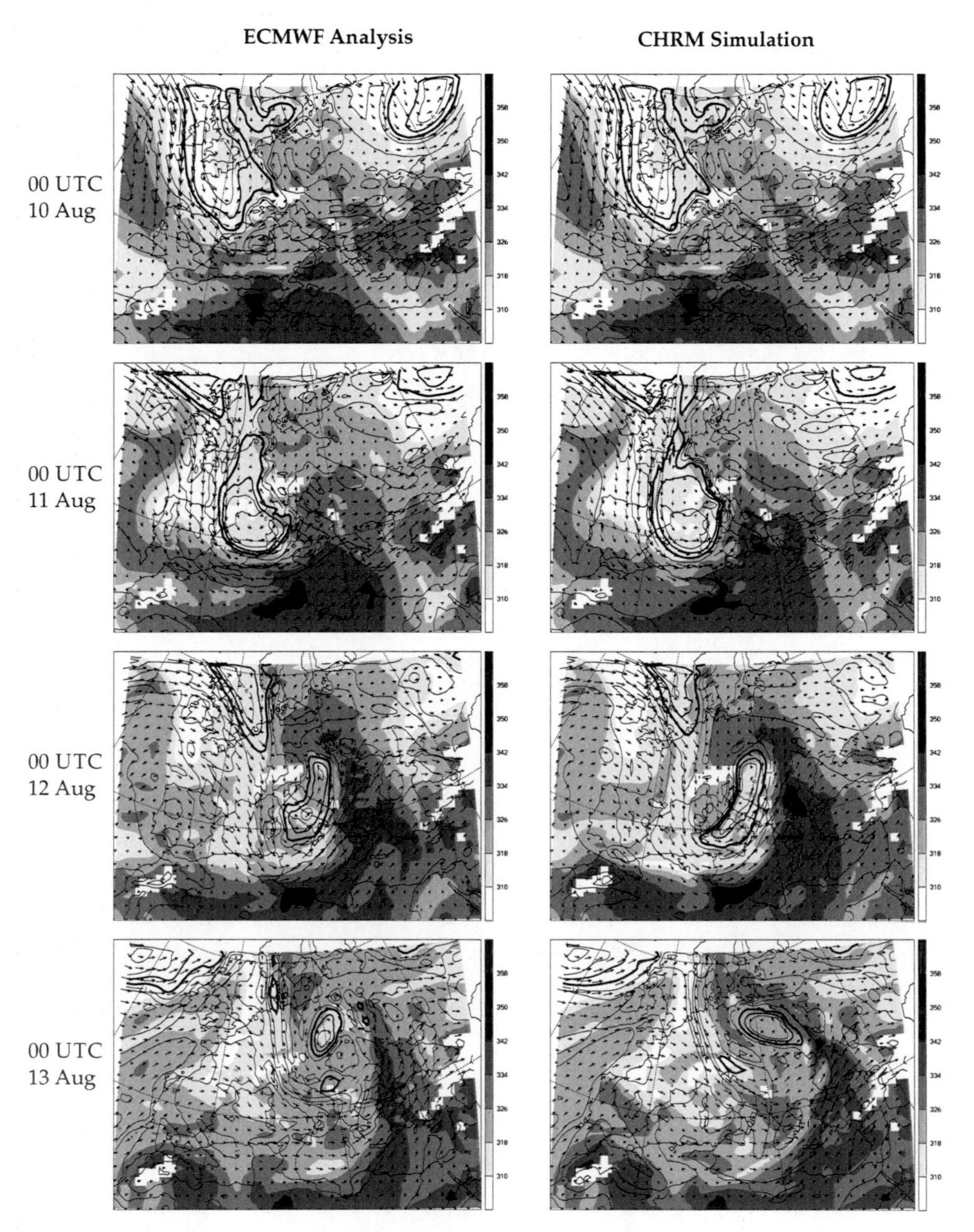

Figure 7.2: Comparison of large-scale circulation features in the upper and lower troposphere between ECMWF analysis data and the CHRM simulation. Contours are potential vorticity (PV) at 325 K, the thick contour denotes the dynamical tropopause at 2 pvu. Shading shows the equivalent potential temperature (θ_e) at 850 hPa in K. Vectors are winds at 250 hPa. Displayed area includes the boundary relaxation zone.

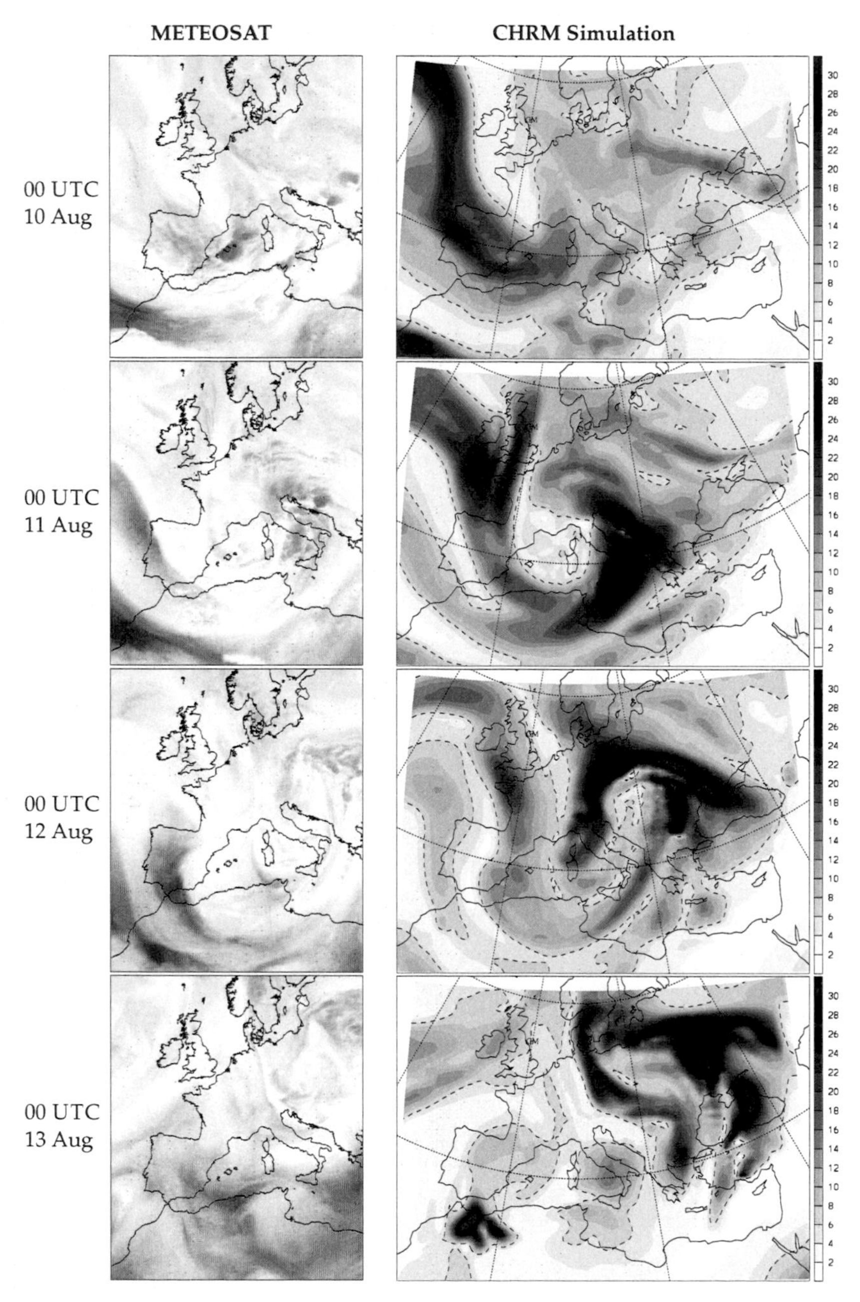

Figure 7.3: Comparison of METEOSAT water vapour imagery (brown=dry, green=moist) with CHRM simulated water vapour, integrated between 150 and 350 hPa. Simulated WV units are 10^{-5} kg kg^{-1}. Dashed contour is at 6×10^{-5} kg kg^{-1}, solid contour at 24×10^{-5} kg kg^{-1}.

Convective contributions are more widespread in, but not limited to, the southern part of the domain. A sequence of convective precipitation maxima is visible on the west coasts of Sardinia, Italy, and the Balkan. With 43%, convective precipitation contributes slightly less than half to the total precipitation in the inner CHRM domain. However, with only 10-20% convective rainfall, precipitation from large-scale processes clearly dominates the maximum over the Erzgebirge. This result is consistent with the model study of Zängl (2004), who found (for a nested simulation down to 1 km horizontal resolution) convective contributions of generally less than 10% over the Erzgebirge, while in total convection made up for about 40% of precipitation in a domain comparable to the one chosen here.

The important role of convective processes[1] for a successful simulation of the Elbe flood is highlighted by a simulation without parametrised convection (not shown). In that case, the missing upward transport of heat and moisture from the boundary layer resulted in an overestimation of surface pressure. In addition, the lack of mid-tropospheric diabatic heating produced less tropopause ridging, which lead to a too far eastward progression of the PV cut-off. As a consequence, the southerly advection of high θ_e air masses at lower tropospheric levels was shut off, and the precipitation maximum displaced far into eastern Europe.

A further simulation with the MPDATA advection scheme for water vapour and cloud water advection showed relatively small differences in the main moisture field,

[1]for the vertical redistribution of mass and energy, rather than for precipitation

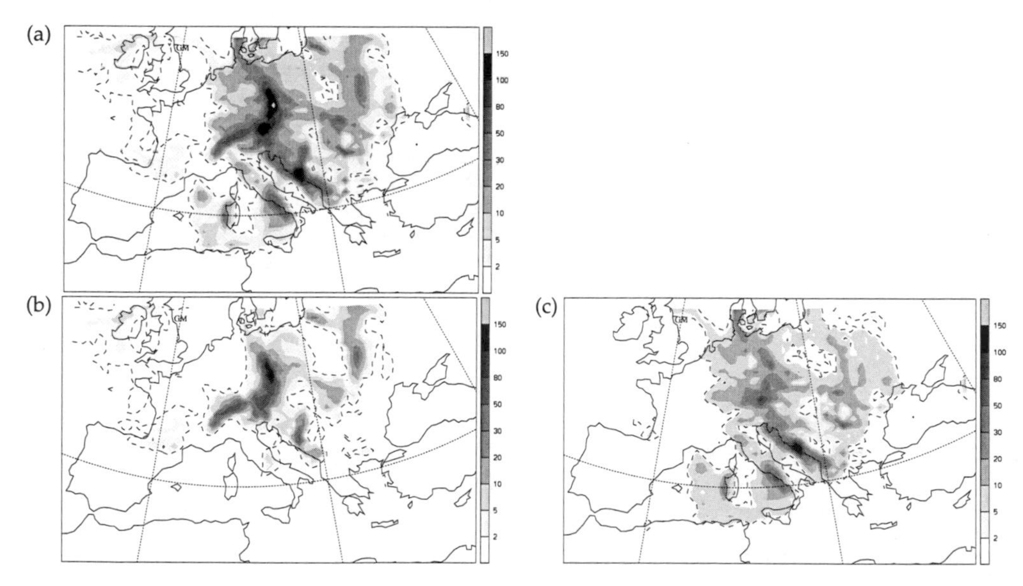

Figure 7.4: Simulated 48 h accumulated precipitation (mm) during 00Z 11 – 00Z 12 August 2002. (a) Total precipitation, (b) large-scale precipitation, (c) convective precipitation. Dashed contour at 1 mm.

but significant deviations with respect to precipitation (not shown). In particular, small precipitation maxima at single grid points (which were probably due to numerical noise from the leap-frog scheme) were no longer present with MPDATA moisture advection. This is a first indication for improved in cloud water advection with the MPDATA scheme, and motivates further experiments and model validation with MPDATA advection.

7.2 WVT experiment setup

The tagging simulations for the August 2002 case study were created with a twofold objective: (i) All relevant moisture sources should be clearly distinguishable with respect to their corresponding moisture transport and contributions to the heavy precipitation in the target area. (ii) The sum of all tagging simulations should ideally comprise all moisture and precipitation in the domain, and hereby allow an assessment of the consistency and representativeness of the method. Hence, six tagging experiments were designed, which in total cover all evaporative and advective sources of water vapour in the model domain (see Fig. 7.5 and Table 7.2):

Atl Atlantic moisture sources; covers the Atlantic ocean area within the domain and atmospheric inflow along the western and northwestern domain boundaries.

Med Mediterranean moisture sources; tags only moisture which evaporates after the start of the simulation (00Z 10 Aug) from the Mediterranean sea within the inner domain.

Bls Black Sea moisture sources; contains Black Sea evaporation and atmospheric inflow along the eastern and north-eastern boundaries of the inner domain.

Sub Subtropical/tropical moisture sources; tags atmospheric inflow along the southern and south-eastern boundaries of the inner domain.

Lnd Land moisture source; contains evapotranspiration from the land surface inside the inner domain after 00Z 10 Aug. This source is important for the assessing the importance of moisture recycling for the Elbe flood.

Atm Atmospheric moisture source; contains the initial ($t = 0$) atmospheric moisture in the inner model domain. This source complements the other initialisations, and thereby allows an assessment of the consistency of the method.

In the following, these six initialisations and the corresponding simulations are, according to the release areas of WVTs, referred to as Atl, Med, Bls, Sub, Lnd, and Atm, respectively. More details on the initialisation procedure with the parameters from Table 7.2 are given in Section 6.1.2.

Table 7.2: Setup of the six WVT experiments for the August 2002 case. Exp: experiment name, Init: initialisation type, t: initialisation time period, x,y,z: grid coordinates for box initialisation, Slab file: name of the slab initialisation file.

Exp	Init	t (s)	x	y	z	Slab file	Description
	Box	0–259200	1–59	58–65	1–27		Atlantic ocean
Atl	Box	0–259200	1–8	1–65	1–27		and boundaries
	Slab	0–259200				slab_atl	above the Atlantic
Med	Slab	0–259200				slab_med	Mediterranean Sea
	Box	0–259200	60–91	58–65	1–27		Black Sea and
Bls	Box	0–259200	84–91	17–57	1–27		eastern European
	Slab	0–259200				slab_bls	boundaries
Sub	Box	0–259200	84–91	9–16	1–27		Subtropical and
	Box	0–259200	9–91	1–8	1–27		tropical boundaries
Lnd	Slab	0–259200				slab_lnd	Land surface areas
Atm	Box	0–0	9–83	9–57	1–27		Initial atmospheric moisture

In a RCM the boundaries pose particular difficulties for the tagging setup. In this particular flow configuration, in the Atl experiment moisture be marked as Atl inflow that crosses the southern boundary domain and then recurses back into the inner domain, while the same moisture in the Sub experiment could be initialised in the boundary zone and enter the domain as Sub tracer. The possibility of such double initialisation will be considered in Sec. 7.3.3.

7.3 Tagging simulations

First, the results of the identification of the moisture transport characteristics and the precipitation for the six tagging experiments are considered, then a first attempt to validate the method with respect to consistency and representativity is undertaken.

7.3.1 Tagged moisture transport

Transport of tagged moisture from the six experiments is displayed in Fig. 7.6 as the averaged tagged specific humidity in the lower troposphere (between the surface and 700 hPa). The 12 h-ly time sequence also shows the cloud water content in the lower

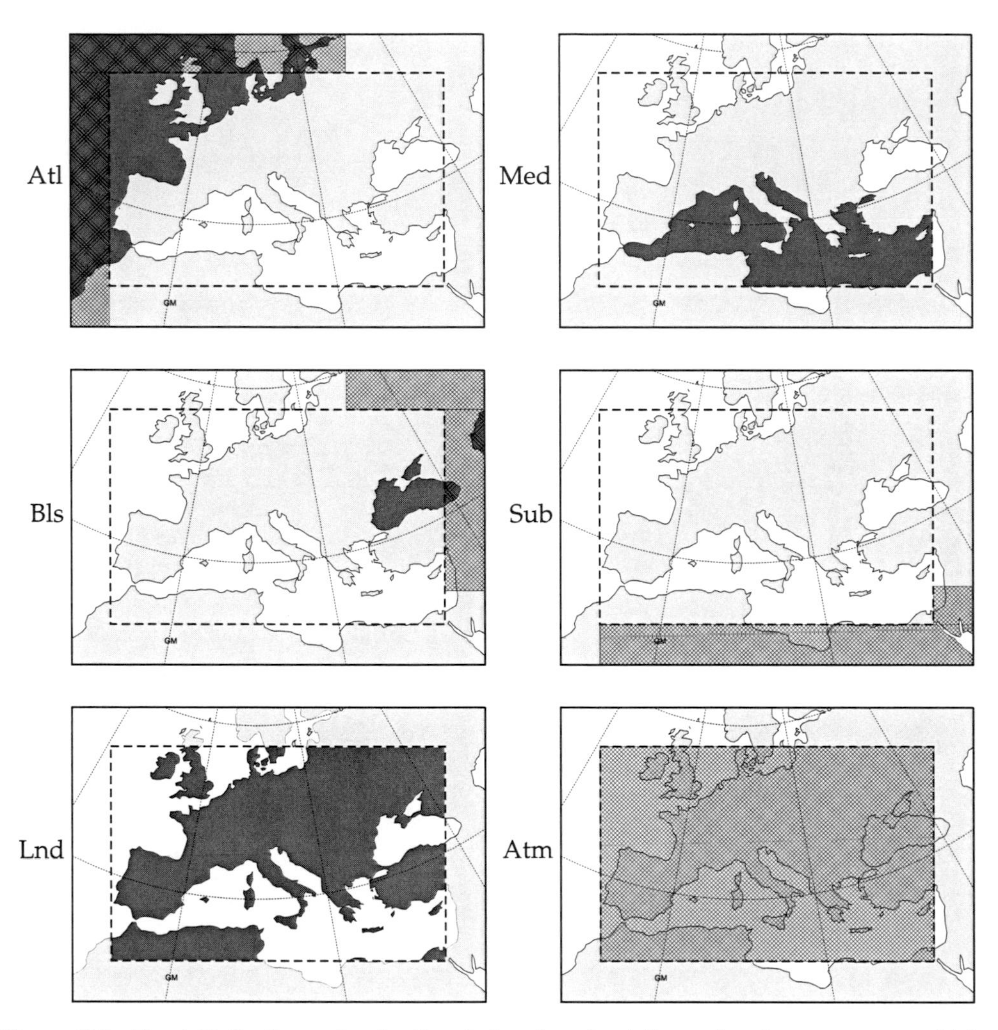

Figure 7.5: The initialisation setup for the six tagging simulations. Land or sea areas which evaporate tagged water (by slab initialisation) are shown in dark gray shading. Atmospheric sections which are initialised with tagged water (by box initialisation) are hatched. The dashed box denotes the area not affected by boundary relaxation. All six initialisations in combination tag all atmospheric water in the model.

troposphere, which should give a first indication of tagged precipitation at a particular instant of time. Salient features of the six experiments can be summarised as follows (a quantitative analysis will be given below):

- Experiment **Atl** shows cyclonic inflow of tagged moisture along the western boundary. Widespread clouds form from that water over the Atlantic, and later orographic condensation/precipitation occurs along the Pyrenees and Alps (Fig. 7.6, 00Z 12 Aug). Some tagged cloud water from the Atlantic also enters the target area during 12Z 12 Aug – 00Z 13 Aug. At the end of the simulation, Atlantic moisture is rather widely dispersed, and dominates in the western part of the domain.

- Experiment **Med** shows a slower dispersion of the tagged water compared to the Atlantic simulation. Recall however that the Mediterranean moisture solely enters the domain from evaporation at the sea surface. Contributions to cloud water in the target area are small, and only appear at 12Z 12 Aug. The bulk of the Mediterranean moisture that evaporates after 00Z 10 Aug is transported into eastern Europe. A clear separation between Mediterranean and Atlantic air masses is visible over central Europe on 00Z 13 Aug. This boundary coincides with the convergence line apparent in Fig. 7.3.

- Experiment **Sub** features an impressive arc-shaped moisture plume that during the simulation extends cyclonically from the subtropical boundary far into eastern Europe. Some clouds form over the Baltic states from this moisture during 12Z 12 Aug to 00Z 13 Aug. No contribution to the area of maximum precipitation is apparent.

- Experiment **Bls** exhibits a slow-moving moisture plume that evaporates from the Black Sea area and slowly progresses to the north-west. Evaporation from the Black Sea appears to be fairly strong compared to advection along the boundaries. Some contribution to cloud water over the Baltic states is apparent on 00Z 13 Aug, and a small part feeds into the target area.

- Experiment **Lnd** shows a maximum over central Europe throughout the simulation (e.g. 00Z 11 Aug). Contributions to cloud water are apparent first along the Alpine range (12Z 11 Aug) and later also in the target area (00Z 12 Aug). In an hourly sequence of the plots in Fig. 7.6, the diurnal variations superimposed on the accumulation of tagged water from surface evaporation can be observed (not shown). By the end of the simulation, the land source has become a small but widespread contributor to moisture in the lower troposphere.

- Experiment **Atm** is initially the largest contributor to moisture in the domain. A large amount of atmospheric moisture is visible as a dark plume over central Europe on 12Z 10 Aug. At the end of the simulation (00Z 13 Aug), it still holds a significant share. Contributions of the Atm tracer to cloud water are substantial over the Atlantic, the Alps, and the target area. The initial Atm moisture is most quickly removed in areas where Atlantic and subtropical inflow is strong (12Z 11 Aug). The patterns of the simulations Atl and Sub for the same instant of time fit into the white areas of the Atm moisture like pieces of a puzzle.

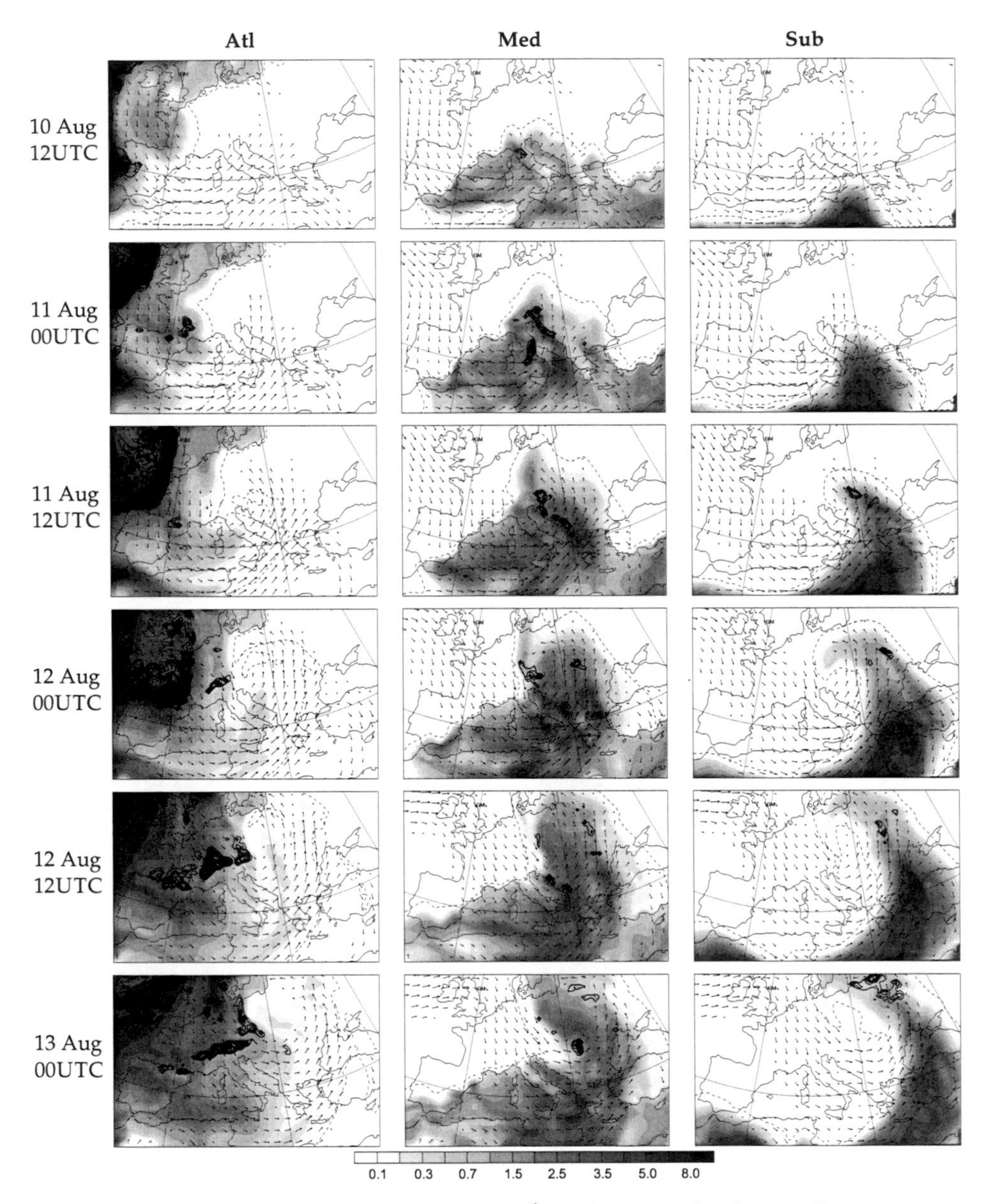

Figure 7.6: Tagged water vapour (shading, g kg^{-1}) and tagged cloud water (heavy contours, contour interval 0.01 g kg^{-1}) averaged between the surface and 700 hPa. Dashed contour indicates 0.01 g kg^{-1} of tagged water vapour. Wind vectors at 700 hPa are shown where the wind velocity is larger than 10 m s^{-1}.

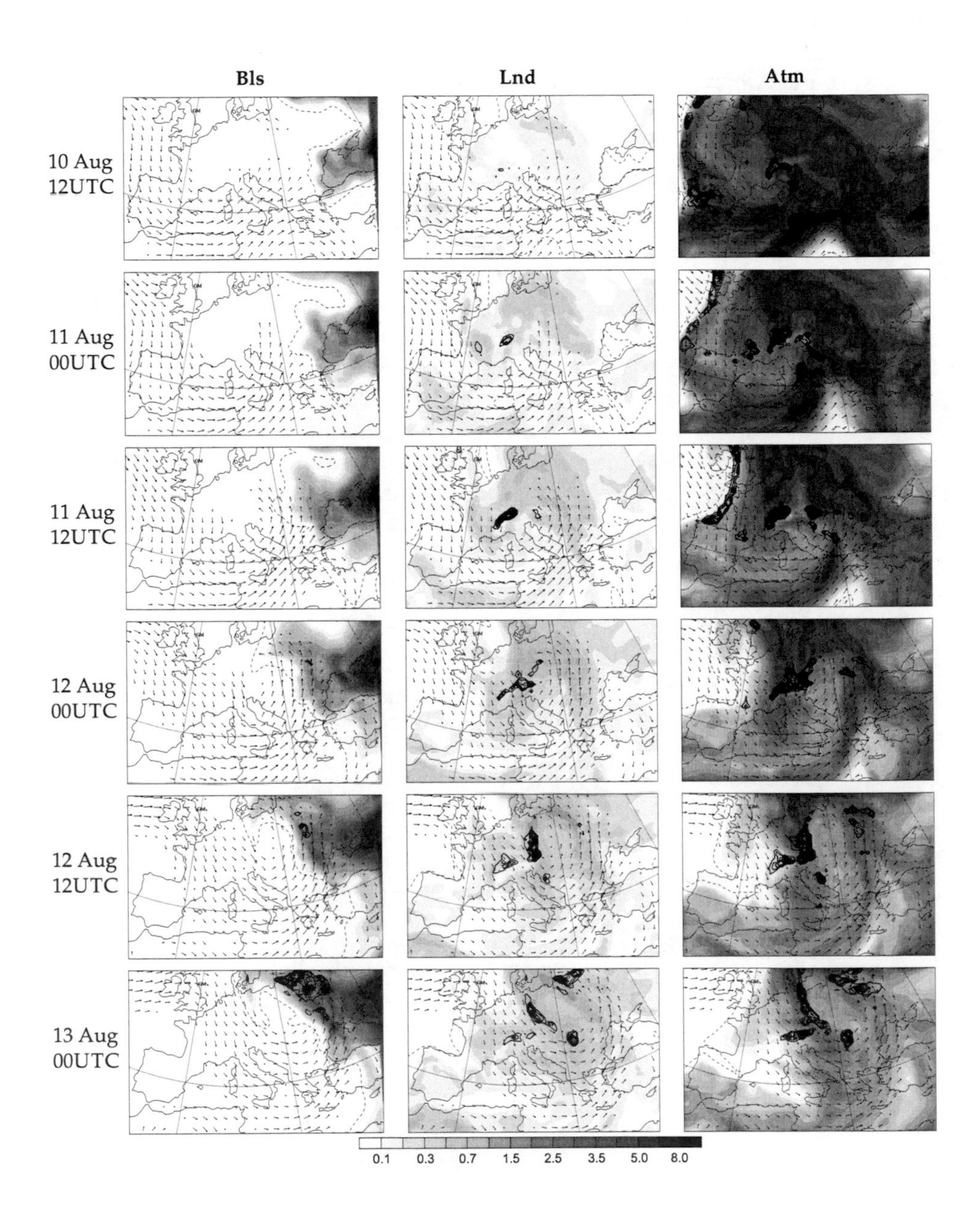

Figure 7.6: Continued.

From the last observation, the general picture emerges that the moisture sources Atl, Med, Bls, and Sub complement each other as local sources with little overlap, while the tracers Lnd and Atm broadly fill the remaining gaps. This shows that the design of the experiments works as intended. A more quantitative investigation of this impression is given in Sec. 7.3.3.

A time series of the total tagged moisture in the inner domain for each experiment gives insight into the strength of each source and the respective time scales (Fig. 7.7a). As the total moisture content in the domain remains fairly constant at 20000 kg m^{-2}, the figure can to a first approximation be interpreted as absolute water masses. For cloud water, the same approximation does not hold. The atmospheric moisture (Atm) which constitutes 100% of the moisture in the inner domain at $t = 0$ is gradually replaced by contributions from the other source areas. At the end of the simulation, about 16.9% of the initial moisture still remain in the domain. The Mediterranean (Med) moisture source gains a share of 18.9% over the 72 h of simulation. Considering that surface evaporation is the only moisture supply, the Mediterranean is a rather strong source region. A further important moisture source is the Atlantic, which gains a share of

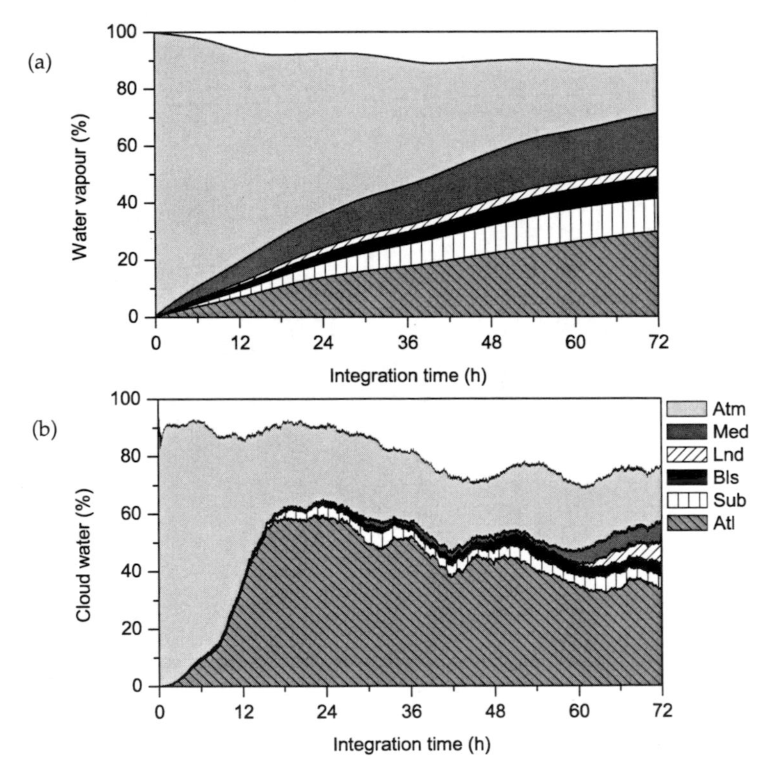

Figure 7.7: Percent contribution of the different moisture sources to (a) water vapour and (b) cloud water in the inner domain. The open area at the top designates moisture which is not assigned to any tracer.

29.8% until the end of the simulation. Unlike the Mediterranean, however, the Atlantic source has a significant contribution from atmospheric inflow at the boundaries. Contributions from the Black Sea (Bls) and the Subtropics (Sub) are comparatively small with about 7.6% and 11.5%, respectively. The land area only contributes a small share (3.6%), but interestingly shows clear diurnal variability. The remaining gap between the sum of all tagged moisture and 100% is discussed in Section 7.3.3. It would be an interesting experiment to run this setup for a longer time (several months) to see if a typical mean moisture composition for this domain can be established.

7.3.2 Tagged precipitation

The contribution of each moisture source to the total 48 h accumulated precipitation from 00Z 11 Aug until 00Z 13 Aug in the inner domain is shown in Fig. 7.8 (compare also Fig. 7.4). In general, tagged precipitation areas reflect the advection paths of the corresponding tagged air masses. The Atlantic (Fig. 7.8, Atl) is (apart from some Atm precipitation) the single contributor to precipitation over the Atlantic sector. Contribution decreases towards the east, and is almost negligible near the target area. The Med experiment (Fig. 7.8, Med) contributes to 40-60% of the precipitation along a broad swath extending from the Mediterranean into eastern Europe, but only just straddles the target area ($< 5\%$). The subtropical source complements total precipitation in a clearly localised sector. The precipitation sum is however small in these regions (Fig. 7.4). The land source (Lnd) has widely dispersed, small contributions to most precipitation areas, with a maximum ($< 10\%$) over central Europe. The Black Sea (experiment Bls) has again very localised contributions to the total precipitation in eastern Europe (~10–15%), and almost none to the target area. The initial atmospheric moisture (Fig. 7.8, Atm) is the largest individual contributor to the moisture in most of the inner domain, and also in the target area (40-80%).

A clearer view of the precipitation contributions to the region of most intense precipitation is derived from a time series of the averaged precipitation rates in the target area for the different experiments (Fig. 7.9a). Two precipitation periods are apparent, one from 00Z 10 Aug to 05Z 11 Aug and a second one that begins on 19Z 11 Aug and continues throughout the remainder of the simulation. This latter one is also the one that caused the devastating floods. In a more narrowly defined target region, precipitation rates would well exceed 4 mm h^{-1}, in agreement with observations (James et al. 2004). The Atm tracer clearly dominates precipitation throughout both events. The first period is to 100% caused from the Atm tracer, and only during the second period other sources become visible. The Med tracer appears first on 20Z 11 Aug, followed by Lnd, Sub, and only on 06Z 12 Aug the Atl tracer, indicating a change of air masses. At the end of the simulation, the 72 h sum of all sources other than the Atm tracer amount to only one eighth of the total precipitation in the target area. Note however that in this simulation nothing can be concluded about the evaporative sources of the Atm water tracer.

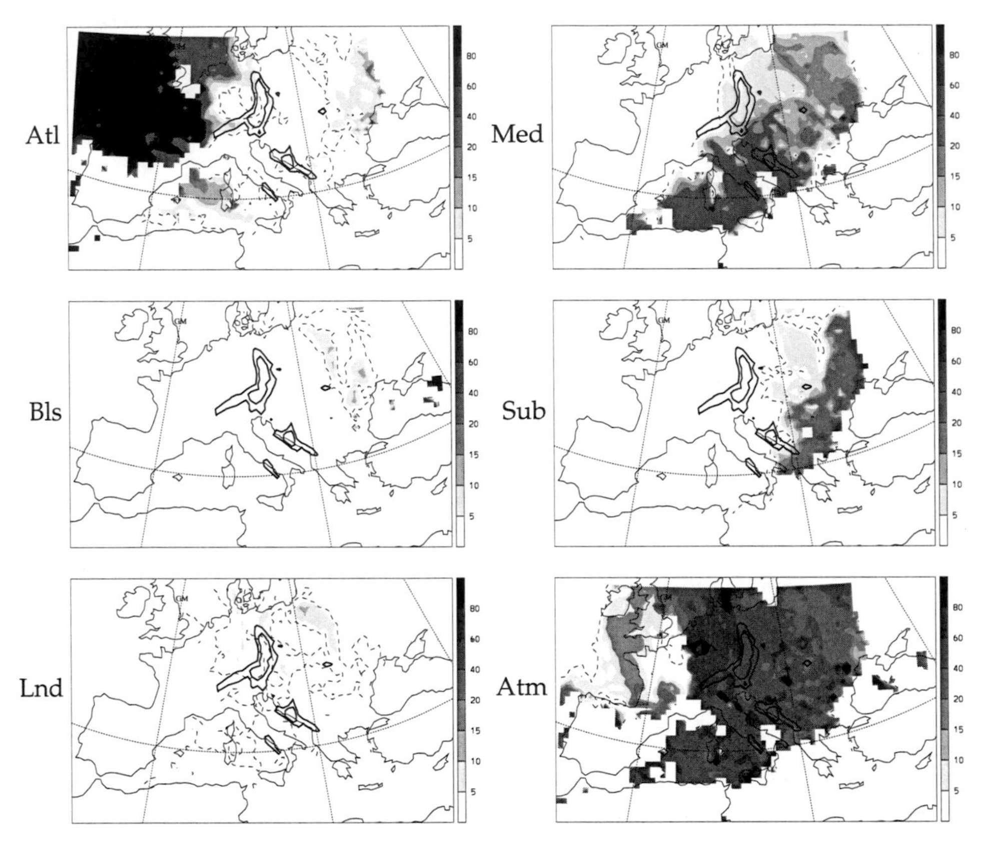

Figure 7.8: Precipitation contributions (shading, % of total precipitation) from the moisture sources corresponding to the six tagging experiments. Dashed line indicates 2%. For reference, the areas with more than 40 and 80 mm 48 h accumulated precipitation during 00Z 11 – 00Z 13 Aug 2002 are indicated by the thick solid contours.

7.3.3 Consistency of the method

The design of the six tagging experiments allows for an assessment of the consistency of the methodology. In an ideal tagging setup, the sum of all tagged moisture contributions should be equal to the total moisture. Fig. 7.10 shows the sum of the relative contribution of each tracer to the total moisture in the lower troposphere. While initially (00Z 10 Aug) all moisture belongs to exactly one tracer (not shown), small inconsistencies develop that lead to excess and unaccounted moisture. An area of excess moisture is visible on 12Z 11 Aug near the southern boundary of the domain that progressively expands into the inner domain. This is probably an effect of re-entering Atlantic moisture as noted on Pg. 155. Some double initialisation also appears near the Black Sea, possibly due to re-entering atmospheric moisture.

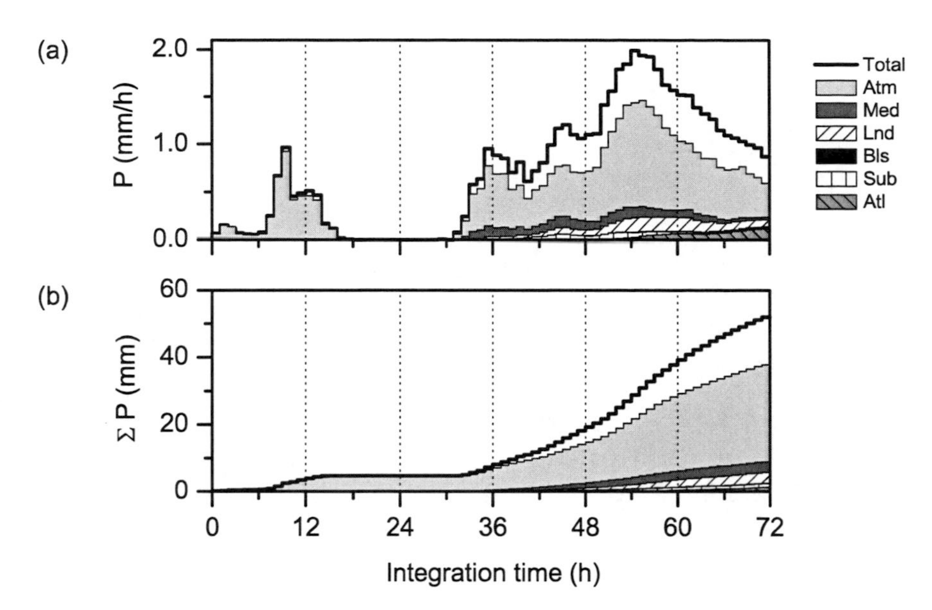

Figure 7.9: Precipitation in the target area from individual moisture sources (shaded) and in total (solid line). (a) Time series of hourly precipitation rates, (b) Accumulated precipitation. The white area between the total and the sum of all tracers denotes moisture not assigned to any WVT.

Unassigned moisture appears in the areas related to the strongest flow convergence, and hence large-scale precipitation formation (e.g. on 00Z 12 Aug in Fig. 7.10 over the target area). By the end of the simulation, less than 60% of the moisture are assigned to a tracer in this area of the domain. As was shown in Section 5.3.2, this behaviour can be expected from the implementation of a synchronised water vapour tracer (STW) in a model that uses non-conservative advection numerics. In convergent flow, the leapfrog advection formulation of the CHRM does not conserve mass, and hence during the synchronisation the tagged moisture is forced to loose mass as well. This loss of mass is hence visible in the areas of strongest convergence in Fig. 7.10, most prominently along the convergence line over central Europe.

Reconsidering Fig. 7.7 gives an impression of the extent of unmarked moisture in the domain. The difference of the total of all moisture sources to the total moisture in the domain is ~5–10% throughout the simulation, and increases to 11.7% after 72 h. As can be gathered from Fig. 7.7b, the STW algorithm affects cloud water more severely than water vapour. Here, the unaccounted moisture amounts to 23% of the cloud water at the end of the simulation. This larger sensitivity could be expected, since the leapfrog advection and MPDATA advection results differ more strongly for a field with sharp gradients, such as cloud water. The stronger sensitivity of the cloud water also leads to a larger fraction of unassigned precipitation.

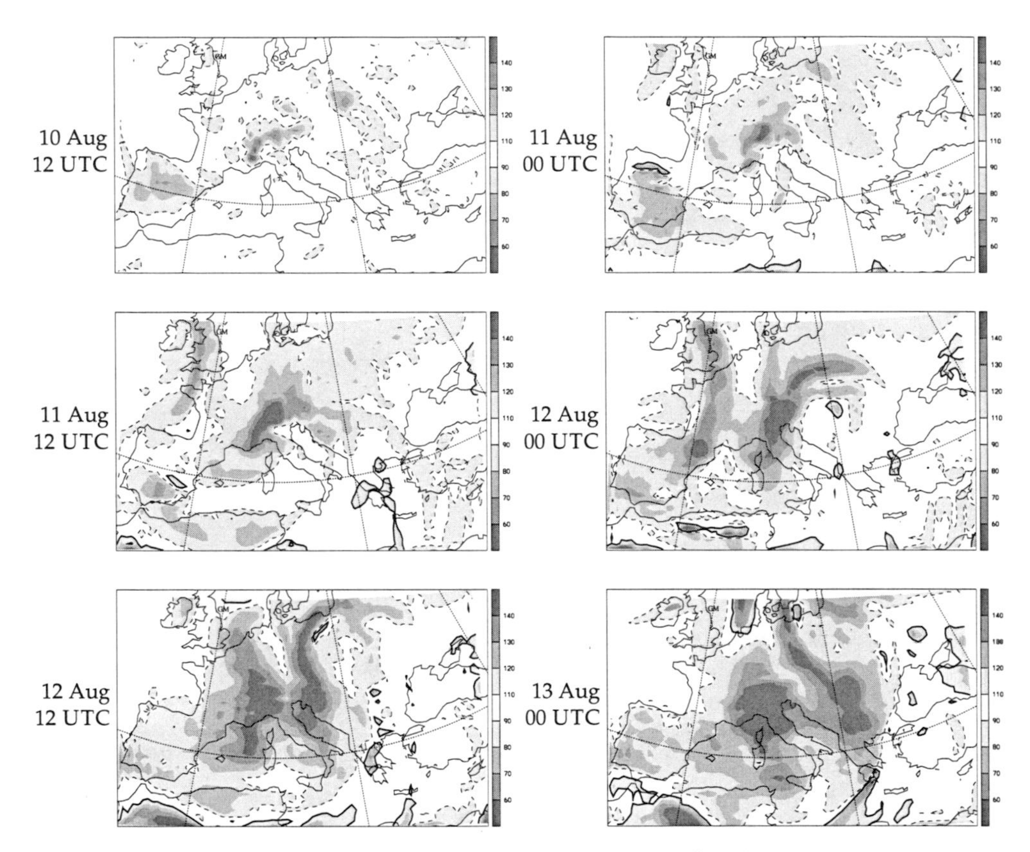

Figure 7.10: Ratio of the total tagged moisture to total moisture (in %) in the lower troposphere (averaged between 700 and 1050 hPa). Areas with unaccounted (overly accounted) moisture are bordered by a dashed (solid) contour.

The degree of inconsistency found here is not unknown to tagging implementations. The tagging studies of Charles et al. (1994), Bosilovich and Schubert (2002), and Numaguti (1999) found inconsistencies of $< 5\%$, 3–5 %, and $< 10\%$, respectively. Most of the gap in the present tagging implementation is probably due to the STW tracer algorithm. As an additional sink of tracer water, the boundary relaxation also reduces the tagged fraction inside the whole domain (see Sec. 6.2.4). The effect is probably not very large ($\sim$2–5%, depending on the distance to the boundary), but may still contribute to the loss of tracer during the simulation.

Furthermore, it is evaluated which fraction of the total precipitation in the domain can be assigned to one of the tracers (Fig. 7.11). The sum of the precipitation from all six experiments accounts for about 80% of the precipitation in the larger part of the domain. Some problematic areas where less than 60% of the precipitation are accounted for are apparent, in particular in the heavy precipitation areas. Reconsidering Fig. 7.9 gives a closer insight into the temporal evolution of the unaccounted fraction of precipitation. Mostly, more than 80% of precipitation are tagged, during the end of the period and at peak precipitation intensities this fraction drops to around 70%. It is a particular contingency of this case study that the target area which is most severely affected by heavy precipitation also shows the largest effect of tracer loss due to the STW tagging method.

7.4 Discussion and further remarks

The lessons learnt from this initial example, both for the implementation of WVT in the CHRM model, and the case study of the Elbe flood, allows to derive some important conclusions which point to future research avenues.

7.4.1 The CHRM tagging implementation

The STW tagging method, which was implemented in the CHRM for this case study, produced results which were in general consistent with the main fields of moisture and precipitation to a high degree. On average, ~90% of the moisture and ~80% of the precipitation could be attributed to a specific tracer. Inconsistencies were to the most part

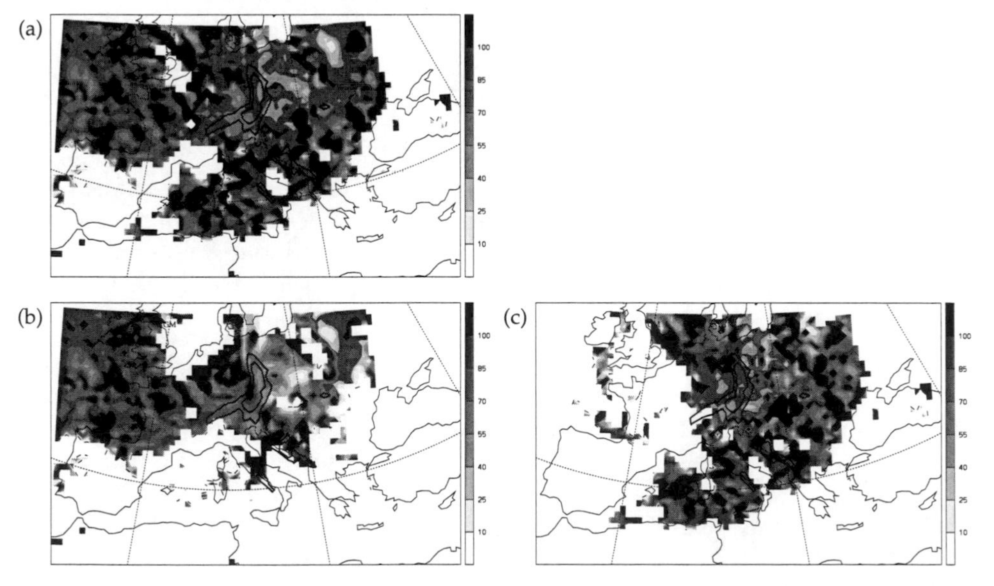

Figure 7.11: Total tagged fraction (in %) of the 48 h accumulated precipitation during 00Z 11 – 00Z 12 August 2002. (a) Total precipitation, (b) large-scale precipitation, (c) convective precipitation.

due to the non-conservative properties of the leap-frog advection in CHRM in convergent flow, which in combination with a synchronised tracer algorithm leads to the loss of tagging information. Using a parallel tracer method (PTW) would have avoided the loss of mass, but comparability with total moisture (in particular the precipitation field) would not have been possible to the same degree as with the STW approach.

Two options are at hand to reduce the fraction of unassigned moisture: (i) One could set the initialisation of the tracer to only 50% of the moisture in a grid cell, instead of 100%, and then multiply the resulting precipitation by two. This would greatly reduce the likelihood of overshoots in convergent areas. The advantage is that the method could be applied as is. However, cloud water overshoots are likely to remain high. (ii) The degree of numerical consistency could greatly be increased by also using the MPDATA advection scheme for the advection of the full water vapour and cloud water fields. This option seems more promising on longer terms. The only caveat here is that such a hybrid version of the CHRM model would at first require new model validation, at least to some degree.

7.4.2 The case study of the Elbe flood

The tagging experiments have shown that moisture contained in the atmosphere at the time of initialisation (00Z 10 Aug 2002) was the single dominant contributor to precipitation over the headwaters of the Elbe river. On the one hand, this finding precludes a full view on the moisture origin of the Elbe flood, which may seem a bit disappointing if one is interested in moisture sources. On the other hand, this finding allows to take a different view on the question of moisture origin. Three days in advance of the flood event, the atmosphere over Europe was already heavily laden with moisture, as can be expected during that time of the year (James et al. 2004). The concurrent evolution of the large-scale flow, governed by an unusually far southward extending stratospheric PV streamer lead to a strongly convergent flow along a quasi-stationary convergence line over central Europe. All the moisture in the atmosphere was then, in analogy to a large wet sponge that was squeezed out horizontally, deposited over a quite limited area, and led to the devastating floods of August 2002.

In that view, the typical 'Vb' cyclone track or even the surface low pressure system were less a reason for the flood as was the large-scale flow. This implies that the quality of the global forecast is important, as it steers the location of flow convergence (comp. Fig. 7.3d), and high-resolution forecasts are important to accurately represent local amplification factors, such as the orography of the Erzgebirge. This view is consistent with the finding of Zängl (2004), who deemed the low quality of the GME analysis from DWD as the reason for the failed forecasts of the Elbe flood with the Lokal Modell (LM) at that time. A comprehensive approach should hence encompass a wider perspective, including besides the influence of moisture sources also the atmospheric features which lead to exceptional large-scale flow patterns.

Despite the dominance of the Atm tracer in the simulations, some conclusions with respect to moisture sources can be drawn. The Atlantic and the Mediterranean are the strongest moisture sources in the target area. Considering the short spin-up time, the evaporation strength of the Mediterranean is particularly noteworthy. A mixture of air masses contributed to the heavy precipitation, which appear in sequence rather than at the same time. This finding compares well with the analysis of James et al. (2004), who found, based on backward trajectories, earlier (06Z 12 Aug) contributions to originate from the Mediterranean, while later (21Z 12 Aug) moisture seemed to originate from the Black Sea and the eastern European land mass. From a Lagrangian moisture diagnostic, Stohl and James (2004) found a strong contribution of the Mediterranean to the precipitation maximum over central Europe. However, their method does not account for rainout during advection, and hence the decreasing importance of further remote moisture sources. This is also an important difference to the Lagrangian moisture source diagnostic developed in this work (see Sec. 3.2). One can only speculate about the moisture origin of the Atm tracer. However, as a first attempt to clarify the moisture origin, of the Atm mass, this tracer initialisation could be divided into several segments, either vertically or horizontally. In the end, however, only a longer integration with a spin-up period of more than three days could provide clearer answers.

Further conclusions can be drawn with respect to the importance of various physical processes for the Elbe flood. Evaporation from sea surfaces with enhanced SSTs is unlikely to have played an important role, at least on a time scale of less than three days in advance of the heavy precipitation. As most moisture was then already in the atmosphere, the high moisture capacity of the atmosphere at that time of the year appears to be more important in that respect. Convective processes, even though apparent in large parts of the domain during the simulation, were not the main cause of precipitation over the target area. It may however have played an important role for the vertical redistribution of moisture and hence to the high moisture content of the middle troposphere, in particular over the Mediterranean. Recycling of soil water or water from the interception storage did play a small, but noticeable role for the precipitation maximum over central Europe. For the actual flood event however, it was probably more important that previous rainfall during 6/7 August had filled the soil water storage, and most of the intense precipitation during 11/12 August went directly into the river runoff.

The Eulerian tagging study with a regional climate model has its particular strengths in representing the various physical processes which redistribute moisture in the atmosphere. Limitations however exist at the temporal and spatial boundaries of such a model. For example, it is an open question if the moisture that entered the domain along the southern boundary (Sub) originated from the subtropical Atlantic, or from the tropics. The Lagrangian tagging diagnostics may be able to complement the Eulerian analysis in such aspects, as it opens a larger spatial domain to the identification of moisture sources. Such aspects of the two methods are investigated further in the following final chapter.

Chapter 8

Final Remarks

Enhancing the understanding of the atmospheric branch of the global and regional hydrological cycle is in the focus of this work. Two new approaches were explored, one Lagrangian and one Eulerian, which allow for the identification of the sources of water vapour for precipitation. Such knowledge holds the promise of a more direct inference of the mechanisms which govern tropospheric water transport. By means of the two methods, precipitation origin has been studied in two target areas of different temporal and spatial scale. In this concluding chapter, the main scientific results and new perspectives from the two applications are summarised first. Then, as this work comprises a strong methodological component, the new findings related to each method development are at first summarised separately, before an attempt to combine the perspectives of Eulerian and Lagrangian thinking with respect to tropospheric moisture transport concludes this thesis.

Inter-annual variability of the source regions for winter precipitation in Greenland

In a Lagrangian analysis of the precipitation origin for the Greenland plateau during selected winter months, the North Atlantic was identified as the sole important moisture source. Source regions of moisture were found to vary strongly with the North Atlantic Oscillation (NAO). An observed shift from sources in the Denmark Strait and Norwegian Sea during NAO positive months to a maximum in the south-eastern North Atlantic for NAO negative months is consistent with differences in the mean large-scale flow between NAO phases. This finding is a new aspect of the influence that large-scale climate modes can impose on the hydrological regime in the target area. An investigation of the transport history of air before moisture content increased over the sea surface pointed towards a crucial role of the land-sea contrast on evaporation and moisture transport processes. A detailed investigation of the role of mid-latitude cyclones and anticyclones for moisture transport to Greenland is left to the future, while the present work forms a well-suited basis to progress with research in such a direction.

In order to link the diagnosed source regions to observational data, the impact of the identified water transport routes on stable water isotope ratios of precipitation over the Greenland plateau was modelled. The MCIM[1] stable isotope fractionation model, supplied with the diagnosed source and transport conditions of water vapour as well as its initial isotopic composition from the ECHAM4 isotope general circulation model (GCM), simulated a similar degree of NAO variability as observed in ice-core data for three winter seasons from central Greenland. Absolute stable isotope ratios, however, showed a significant lack of depletion compared to ice-core data. The variability with the NAO can be qualitatively explained by combined source and transport influences. As has recently been recognised (Masson-Delmotte et al. 2005a), these results indicate the need for a multi-causal interpretation of stable isotope signals in ice cores. The high spatial resolution of this work could be helpful for determining ice-core sites on the Greenland plateau where isotopic signals should show strong inter-annual variability. In the future, the diagnosed water vapour transport conditions could be used to re-tune the stable isotope model, and then to expand the study period to several full seasonal cycles.

Identification of the source regions for precipitation during the Elbe flood, August 2002

A three-day tagging simulation with the CHRM[2] regional climate model (RCM) of the Elbe flood (10–13 August 2002) was performed. The results highlighted the importance of the concurrent upper-level circulation for producing the extreme precipitation in parts of Germany, Austria, and the Czech Republic during that period. Identification of the role of the most influential evaporative moisture sources was severely limited by spin-up effects. The largest part of the precipitation in the flood area during the model simulation was produced by the moisture that was present in the atmosphere at the time of initialisation. However, different evaporative sources did contribute to the extreme precipitation in the most affected area; notably at distinct, subsequent periods of time. In the future, a simulation over an extended period, starting several days earlier, could provide a more conclusive view with limited additional effort. Such a refined study could then allow to more clearly separate different precipitation-related processes, such as moisture advection from the Mediterranean, and recycling of precipitation from the land surface.

[1] Mixed-Cloud Isotope Model
[2] Climate High-Resolution Model

Lagrangian moisture transport diagnostic

A Lagrangian method has been developed which is able to diagnose the sources and transport paths of water vapour traced along 3-dimensional kinematic back-trajectories. The method considers the full transport history of an air parcel. By taking precipitation and subsequent uptakes *en route* into account, each sources' contribution to diagnosed precipitation at the arrival location can be determined. The method has been expanded progressively, and can now be applied to other hydrological problems. The flexibility of the method has been demonstrated in an application to stable isotope fractionation modelling. On this basis, a further expansion towards a Lagrangian box model of stable isotope fractionation using back-trajectories could be envisaged. Furthermore, with minor adjustments, source-receptor relationships can be derived for substances which only partly participate in the atmospheric hydrological cycle, such as soluble pollutants, or mineral dust (see Appendix A).

Eulerian water vapour tagging in a RCM

Water vapour tagging (WVT) was implemented into the well-verified CHRM model, which is, however, built on non-conservative water advection numerics. For that purpose, an algorithm was developed which assured consistency between the conservative MPDATA[3] tracer advection and the model's water fields at the expense of mass conservation. Numerical consistency, initialisation and spin-up, and boundary treatment proved to be critical points of the WVT implementation and application. A major advantage of this tagging RCM, compared to the currently existing GCM tagging models, is the possibility to use reanalysis data as boundary fields, and hence to conduct case studies which can be compared to observational data. Further completion of the method will require a treatment of the soil model, and testing of fully conservative water advection numerics in the model. The method is then open to a wide range of applications, and in particular suitable for process studies in the regional water cycle. As a very attractive option appears the nesting of the tagging RCM into future climate simulations with a tagging GCM.

Towards a comprehensive picture

Even though the Lagrangian method has been applied to a problem on inter-annual time scales and the Eulerian method for a particular 72 h case study, time is not *a priori* a limitation for either one of the methods. The quality of the used reanalysis data, which is typically better in regions and periods of higher data coverage, is in turn a limitation that is common to both methods. Each method also has individual strengths

[3]multidimensional positive definite advection transport algorithm with non-oscillatory option

and limitations: The results of the RCM tagging depend crucially on the conservative behaviour of the model's numerics, and on the physical plausibility of sub-grid scale parametrisations. The Lagrangian approach is even more affected by the latter processes: Parametrisations of convection and precipitation evaporation, as well as inconsistencies between subsequent reanalysis time steps and trajectory calculation uncertainties can cause fluctuations in the moisture field that are not captured by this method. Unlike the limited calculation domain of a RCM tagging study however, the Lagrangian moisture transport diagnostic can be applied on a global scale, and without spin-up be expanded in time. In addition, information on the transport history of air parcels such as the pre-uptake history and other transport-related parameters are available from this method, which are not (or not directly) accessible with an Eulerian approach.

A direct comparison and assessment of the methods' consistency is hampered by the fact that both provide a different kind of moisture source information: While the Eulerian method is a forward calculation with specified sources and an *a priori* unknown target, the Lagrangian approach used here is a backward calculation from a specified target region with *a priori* unknown sources (Holzer et al. 2005). The source attribution included in the Lagrangian diagnostic developed here provides a detailed structure of the source region; a result that is conceptually identical to an Eulerian tagging simulation with different tracers released from each location at the model surface. The Lagrangian source regions can hence be directly compared to the results of Eulerian WVT methods when the information is binned into larger evaporative tracer source regions, as has been done here for the comparison of diagnosed Greenland moisture sources with the Eulerian tagging results of Charles et al. (1994). An application of the two methods developed here to a single identical case study would therefore make it possible to directly compare the respective results, and to assess their consistency. In particular, the influence of the abovementioned processes that are not captured by the Lagrangian method could be quantified.

The comparison of both method's limitations also suggests that each approach has different scales and problem settings which it should preferentially be applied to. When problems on the regional scale are considered, or parametrised processes, in particular convection, are important, an RCM should give the better results. When large-scale transport and processes clearly dominate over parametrised processes, the Lagrangian diagnostic can reveal moisture origins without being limited by a specific RCM domain and spin-up. In the tagging case study of this work, for example, the Lagrangian diagnostic could be applied to reveal the evaporative sources of moisture entering the model domain along the southern boundary. In this respect, the two methods could be applied jointly, and thereby provide a complementary Lagrangian and Eulerian picture.

In conclusion, with the Lagrangian and Eulerian methods developed in this work, new insight into the characteristics of and the processes related to the atmospheric branch of the hydrological cycle is now in reach.

Mais au Nord, le ciel est ce soir pur de nuages. Et le vent a changé de goût. Il a aussi changé de direction. Nous sommes frôlés déjà par le souffle chaud du désert. C'est le réveil du fauve ! Je le sens qui nous lèche les mains et le visage !

Antoine de Saint-Exupéry, Terre des Hommes

Appendices

Appendix A

The transport history of two Saharan dust events archived in an Alpine ice core

Abstract[1]

Mineral dust from the Saharan desert can be transported across the Mediterranean towards the Alpine region several times a year. When coinciding with snowfall, the dust can be deposited on Alpine glaciers and then appears as yellow or red layers in ice cores. Two such significant dust events were identified in an ice core drilled at the high-accumulation site Piz Zupó in the Swiss Alps (46°22′ N, 9°55′ E, 3850 m a.s.l.). From stable oxygen isotopes and major ion concentrations, the events were approximately dated as October and March 2000. In order to link the dust record in the ice core to the meteorological situation that led to the dust events, a novel methodology based on back-trajectory analysis was developed. It allowed the detailed analysis of the specific meteorologic flow evolution that was associated with Saharan dust transport into the Alps, and the identification of dust sources, atmospheric transport paths, and wet deposition periods for both dust events. Differences in the chemical signature of the two dust events were interpreted with respect to contributions from the dust sources and aerosol scavenging during the transport.

[1]Chapter has been published as H. Sodemann, A. S. Palmer, C. Schwierz, M. Schwikowski, and H. Wernli, *The transport history of two Saharan dust events archived in an Alpine ice core*, Atmospheric Chemistry and Physics, 6, 763-796, 2006.

For the October event, the trajectory analysis indicated that dust deposition took place during 13–15 October 2000. Mobilisation areas of dust were mainly identified in the Algerian and Libyan deserts. A combination of an upper-level potential vorticity streamer and a midlevel jet across Algeria first brought moist Atlantic air and later mixed air from the tropics and Saharan desert across the Mediterranean towards the Alps. The March event consisted of two different deposition phases which took place during 17–19 and 23–25 March 2000. The first phase was associated with an exceptional transport pathway past Iceland and towards the Alps from northerly directions. The second phase was similar to the October event. A significant peak of methanesulphonic acid associated with the March dust event was most likely caused by incorporation of biogenic aerosol while passing through the marine boundary layer of the western Mediterranean during a local phytoplankton bloom. From this study, we conclude that for a detailed understanding of the chemical signal recorded in dust events at Piz Zupó, it is essential to consider the whole transport sequence of mineral aerosol, consisting of dust mobilisation, transport, and deposition at the glacier.

A.1 Introduction

Dust plays a significant role in the global climate system. It influences the radiative properties of the atmosphere, plays a complex role in cloud formation, and is an important source of nutrients in biogeochemical cycles (Prospero and Lamb 2003). The Saharan desert is the most important natural source of dust (Goudie and Middleton 2001). Frequently, Saharan dust is ejected into the Mediterranean atmosphere (Moulin et al. 1998). In some instances, Saharan dust has been transported as far as northern Scandinavia (Franzén et al. 1995). More commonly, Saharan dust may reach central Europe and in particular the Alpine region up to several times a year. Prodi and Fea (1978, 1979) inferred Saharan dust events in the Italian Alps from air filter samples, and identified typical synoptic situations which are associated with this dust transport pathway.

Precipitation that contains scavenged Saharan dust is known as "red rain" or "yellow snow" in some areas. Along with the colour, the dust load strongly imprints upon the precipitation chemistry. Several investigations focused on the identification of the source areas of dust from these chemical characteristics. By analysing "red rain" in northeastern Spain, Avila et al. (1997) detected differences in the mineralogical and soluble ion chemistry composition for different source regions in northern Africa. During a large Saharan dust event in the Alps, Schwikowski et al. (1995) found characteristic mass concentrations of dust-related ions in aerosol and precipitation samples to be consistent with mineralogical indicators and back-trajectories. At higher elevation sites, scavenged Saharan dust may be deposited as snow, buried as distinct layers in firn and ultimately preserved in glaciers. Wagenbach and Geis (1989) used this fact to identify

Saharan dust events in an ice core retrieved from a glacier in the Swiss Alps. Hence, given that the age of the ice can be determined accurately, ice cores can be used as an archive of such Saharan dust events. Conversely, if the characteristics of a particular Saharan dust event are known in detail, it can be used as a reference horizon for the dating of other Alpine ice cores.

The chemical composition of the mineral aerosol can differ largely between individual dust events. Usually, these differences are attributed to characteristics of the various mobilisation areas of dust (Bergametti et al. 1989; Avila et al. 1997; Claquin et al. 1999). However, a dust plume may also be chemically altered during its transport to the deposition site, for example by the uptake of predominantly marine compounds such as sodium (Na^+) and chloride (Cl^-) over the ocean (Schwikowski et al. 1995), or by predominantly anthropogenic pollutants, such as ammonium (NH_4^+), residing in the boundary layer (BL) (Jones et al. 2003). Other aerosols, such as the exclusively marine biogenic compound methanesulphonic acid (MSA, CH_3SO_3H) may even provide information on ocean-atmosphere interaction. It therefore may be essential to consider the complete transport history of dust-laden air parcels to gain a detailed understanding of the chemical signal of a specific dust event in the ice core. This comprises three distinct stages in the life cycle of a dust event, namely dust mobilisation at the source area, dust transport according to the specific meteorological situation, and finally dust deposition at the observation site. All three of these stages are characterised by specific meteorological conditions and processes.

In this study, we reconstruct the transport history of two large dust events which were recorded in an ice core drilled at Piz Zupó in the Swiss Alps. The ice core section attributed to the year 2000 shows two pronounced episodes of enhanced concentrations of dust tracers (e.g. Ca^{2+}). We interpret the chemical signatures of these two dust events by means of an extended back-trajectory analysis. Our investigation focuses on two main questions:

(i) What influences the dust origin, transport pathway and deposition of a chemical signal recorded in an ice core? In other words, are differences in the chemical signal for different dust events due to a differing dust origin, transport or deposition process - or a combination thereof?

(ii) Is there an archetypal meteorological flow evolution that leads to Saharan dust transport to the Alps, or can different types of meteorological events in principle lead to similar dust signals in Alpine ice cores?

A common way to address these two questions are backward trajectories. Back-trajectory analysis has been used as a tool for studying the transport history of dust in a number of studies (e.g. Schwikowski et al. 1995; Avila et al. 1997; Collaud Coen et al. 2004; Bonasoni et al. 2004; Barkan et al. 2005; Papayannis et al. 2005). Most of these studies, however, inferred dust source regions directly from the horizontal position of an air parcel several days before arrival, without taking into account its vertical position or any further information. Here, we propose a new methodology which is based on back-trajectories, yet in addition makes use of meteorological information along the air parcels' flight paths. By means of objective criteria, locations are extracted where dust mobilisation, chemical interaction, and wet deposition are likely to occur during dust transport. This enhanced back-trajectory method provides a closer link between (i) the meteorological processes leading to mobilisation, transport, and deposition of a dust plume, and (ii) the chemical signal at the deposition site, than would be possible with the use of backward trajectories alone. This approach can be considered as a first step towards explicit Lagrangian dust transport modelling.

With this study, we aim to enhance the understanding of the chemical fingerprints left in the ice core by specific meteorological processes. Today's high-quality meteorological data and satellite observations provide an ideal opportunity to investigate how recent Saharan dust events become preserved in an ice core. Ultimately, a better understanding of the relevant processes could provide important keys on how to interpret whole ice core records as archives of the frequency and amplitudes of past Saharan dust events.

A.2 Data

A.2.1 Ice core site and analysis

In March 2002, a 43 m long ice core, corresponding to ~29 m water equivalent (w.e.), was retrieved from the high-altitude site Piz Zupó (46°22′ N, 9°55′ E, 3850 m a.s.l.) in the Swiss Alps (Fig. A.1). The drilling site was located on the saddle between Piz Zupó and Piz Argient, and was exposed to advection from north-westerly and south-easterly directions. Despite being on a saddle, with 2.6 ± 0.8 m w.e. for the time period 1992-2001 the mean annual accumulation rate at the drilling location is rather high (Palmer et al., 2005[2], see also Sec. A.3). Mean annual precipitation from nearby lower-elevation meteorological stations is much lower (0.93-1.20 m) for the same period of time, while the year-to-year variations agree well. This fact together with the high accumulation rate at the ice core site indicates that wind erosion is not significant at the drilling site.

The ice core was examined in the freezer (−20 °C) at the Paul Scherrer Institute prior to decontamination procedures. Length, diameter, density and visual features (dust layers and ice lenses) of the individual core sections were noted. Ice core sections were then cut into 4–5 cm sections using established techniques (Eichler et al. 2000).

The samples were analysed for a suite of anions and cations (F^-, CH_3COO^-, $CHOO^-$, MSA, Cl^-, NO_3^-, SO_4^{2-}, $C_2O_4^{2-}$, Na^+, NH_4^+, K^+, Mg^{2+}, Ca^{2+}) using ion chromatography (IC). Cations were measured using a 16 minute isocratic method with 20 mM MSA eluent at 1 mL min^{-1}. Anions were analysed in 19 minute method involv-

[2]Palmer, A. S., Jenk, T., Schwikowski, M., Saurer, M., Schwerzmann, A., Lüthi, M., Funk, M., and Gäggeler, H. W.: A new ice core record from Piz Zupó, South-East Switzerland, in preparation, 2005.

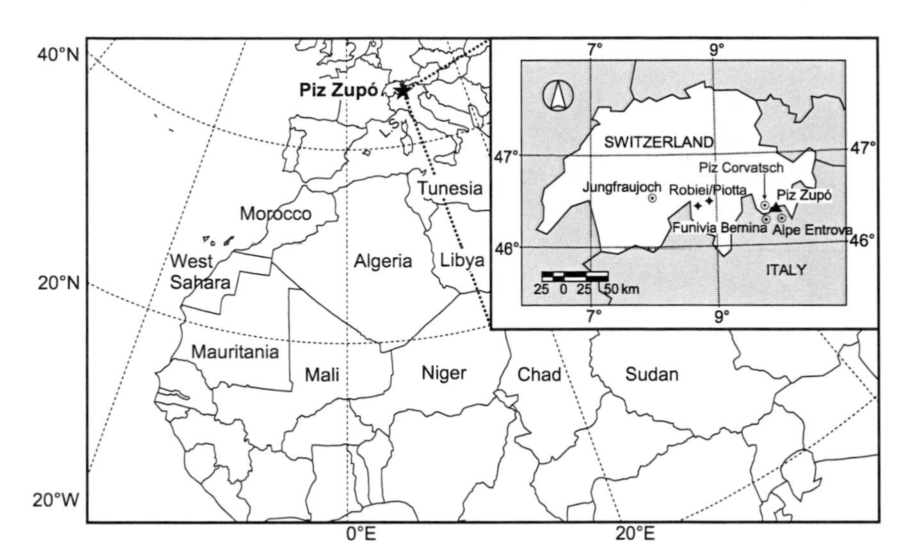

Figure A.1: Geographical map of the countries containing parts of the Saharan desert. Inset shows the location of the ice core site (Piz Zupó), the aerosol measurement site (Jungfraujoch), two stations where precipitation chemistry is analysed at irregular intervals (Robiei, Piotta), and three precipitation stations near Piz Zupó (Piz Corvatsch, Funivia Bernina, Alpe Entrova). The Ligurian Sea is denoted by L. S.

ing gradient elution (concentrations between 0.25 to 24 mM NaOH) at 0.5 mL min^{-1}. Suppressed conductivity detection was used in both IC methods. Measurement of the oxygen isotope ratio, $\delta^{18}O$ (defined as the relative deviation of the $^{18}O/^{16}O$ ratio of the sample from the international standard VSMOW) at Paul Scherrer Institute involved the pyrolysis of water to CO at 1450°C in a glassy carbon reactor (Saurer et al. 1998; ornexl et al. 1999), and subsequent mass-spectrometric analysis of $^{12}C^{16}O$ and $^{12}C^{18}O$ (Delta Plus XL, Finnigan MAT, Bremen, Germany). A complete description of the procedures used for decontamination, analysis and dating of the Piz Zupó ice core are given in Palmer et al. (2005).

A.2.2 Precipitation data

Precipitation data from three automated weather stations located in the vicinity of Piz Zupó were used as observational indicators of wet deposition at the glacier site (Fig. A.1). The Swiss station Piz Corvatsch (46°25′ N, 9°49′ E, 3015 m a.s.l.) is located on a ridge ~10 km northwest of Piz Zupó and ~800 m lower than the ice-core drilling site. At this altitude, reliable precipitation measurements are difficult due to frequent snowfall, and some rain shadowing during southerly flow may occur. Still, with respect to timing and intensity, the precipitation registered here should compare reasonably with the precipitation at Piz Zupó itself. The two stations P41 and P42 are located at Alpe Entrova (1905 m a.s.l.) and Funivia Bernina (2014 m a.s.l.), respectively, on the southern

(Italian) slope of the Bernina massif. These weather stations are likely to be influenced by orographic shadowing and congestion, and hence can show different precipitation magnitudes and timings to Piz Zupò, depending on the flow direction. All half-hourly precipitation data were aggregated into 6 h periods centered around 00Z, 06Z, 12Z, and 18Z for better comparison with the other meteorological data.

A.2.3 Satellite imagery

The concentration of chlorophyll *a* (chlo-*a*) at the sea surface is a proxy for the biogenic contribution to organic aerosol in the marine BL (O'Dowd et al. 2004), which via oxidative processes also includes MSA (Huebert et al. 2004). Hence, in this study we used composite Sea-viewing Wide Field-of-view Sensor (SeaWiFS) chlo-*a* products processed with the OC4v4 algorithm. The OC4 algorithm version 4 was proposed by O'Reilly et al. (1998). It provides the combined chlo-*a* and pheophytin *a* concentrations from an empirical relation to the maximum remote-sensing reflectance ratio at the sea surface with respect to several selected wavelengths.

One-week composites of chlo-*a* averaged over the periods 07–14 October and 13–20 March 2000 were acquired from the SeaWiFS Project Homepage (see http://oceancolor.gsfc.nasa.gov/cgi/level3.pl). Gaps in the weekly composite maps due to clouds were filled with the respective monthly mean values. The OC4v4 algorithm is known to systematically overestimate chlo-*a* concentrations in oligotrophic areas in the Mediterranean (Bosc et al. 2004). Therefore, we applied the regional bio-optical algorithm proposed by Bricaud et al. (2002) to the chlo-*a* maps of both periods, which by means of an empirical relationship decreases OC4v4 chlorophyll *a* concentrations below 0.4 mg m^{-3} by a factor of up to 2.5.

In addition, visible SeaWiFS imagery was acquired for the visual identification of dust plumes leaving the African continent. Finally, half-hourly Meteosat infrared (IR) imagery processed by MeteoSwiss were used for the inspection of cloud patterns and potential precipitation areas during the two study periods.

A.2.4 Meteorological data

ERA-40 reanalysis data (Simmonds and Gibson 2000) from the European Centre for Medium-range Weather Forecast (ECMWF) were the basis of the meteorological analysis and the back-trajectory calculations. The data were used on the original model levels and interpolated onto a regular grid at $1^\circ \times 1^\circ$ resolution. Considered primary variables include horizontal and vertical wind velocities, temperature, and specific humidity. From these quantities, secondary variables like sea level pressure, potential temperature and potential vorticity were calculated.

The meteorological situation determining the transport of dust parcels was examined from time series of maps of various meteorological fields. Upper-level potential vorticity (PV) on the 320 K isentropic surface was used to identify the larger-scale synoptic situation. Horizontal wind velocities at 200 and 500 hPa served to highlight influences of upper-level jets on dust mobilisation and transport. Also considered were the vertically averaged horizontal moisture flux

$$F = \frac{1}{p_s - p_t} \int_{p_t}^{p_s} q \cdot |\vec{v}| dp \tag{A.1}$$

(with q being the mixing ratio in $g\,kg^{-1}$, $\vec{v}$ the horizontal wind vector) between pressure level p_t=700 hPa and the surface (p_s), and wind vectors at 700 hPa. Finally, sea level pressure (SLP) and equivalent potential temperature (θ_e) at 850 hPa allowed identification of the synoptic situation and airmass differences at lower levels. Vertical cross-sections of these meteorological fields together with the positions of the identified dust trajectories were useful to investigate potential interactions between dust clouds and the marine BL at selected times.

A.3 Ice core chemistry of dust events

Dating the ice core by annual layer counting was started at the surface which corresponds to the date of drilling (March 2002), making the attribution of the year 2000 relatively straightforward. The ice core's $\delta^{18}O$ and chemical records, in particular NH_4^+, show clear seasonal variations with high values during summer and low values in winter, allowing for annual dating of the core (Eichler et al. 2000; Preunkert et al. 2000). This dating implied that the Piz Zupó ice core spans the time period 1991–2001 ($\pm$1 yr). The relatively large annual accumulation (2.6 m w.e.) allowed for a high resolution analysis of the ice core chemistry. The sampling scheme was chosen to provide at least 12 samples per accumulation year. In the case of the core section described here, a much higher resolution due to the high accumulation rate results. As the section ascribed to the year 2000 consists of about 100 samples, the time resolution per sample probably ranges between several hours to weeks, depending on the frequency and intensity of precipitation events at the site.

Initial study of the Piz Zupó record for the year 2000 showed two large Ca^{2+} peaks, typical indicators of mineral dust transport to the site (Fig. A.2). A number of smaller excursions in the Ca^{2+} record were observed between the two large events, but are not further examined here.

The first (later) large dust event was located at a depth of 4.5–4.7 m w.e. in the ice core and had high concentrations of Na^+, K^+, Mg^{2+}, Ca^{2+}, Cl^- and SO_4^{2-}. This event shows virtually no MSA and low concentrations of NH_4^+ and NO_3^-. The isotopic and chemical signatures before and after this event were characteristic of early winter conditions following the decrease of anthropogenic input in summer (Eichler et al. 2000; Preunkert et al. 2000). Yellow/brown particles were observed in the ice samples for this section of the core. There are three reports of dust deposition in Southern Switzerland

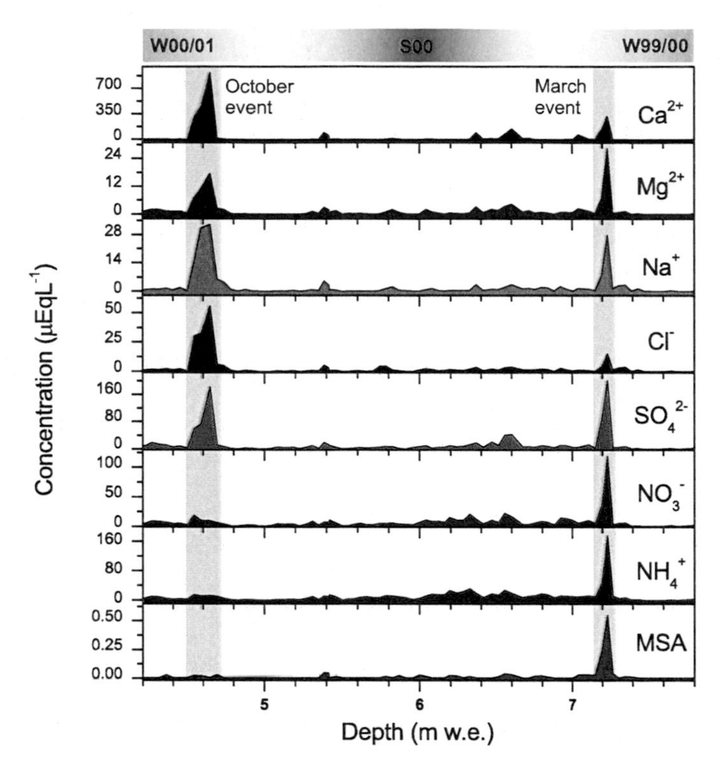

Figure A.2: Chemistry in the ice core from Piz Zupó for the year 2000. Shown as a function of ice core depth in m water equivalent (w.e.) are the concentrations of calcium (Ca^{2+}), magnesium (Mg^{2+}), sodium (Na^{+}), chloride (Cl^{-}), sulfate (SO_4^{2-}), nitrate (NO_3^{-}), ammonium (NH_4^{+}), and methanesulphonic acid (MSA) in micro equivalent per litre (μ Eq L^{-1}). H^{+} (not shown) reflects the acidity of the ice and behaves very similar to NO_3^{-}, while potassium (K^{+}) and oxalate ($C_2O_4^{-}$) (not shown) peak similarly to Mg^{2+}. The areas shaded in gray denote the sections influenced by large amounts of mineral aerosol (dust events).

during 13–15 October 2000. (1) Observation of "red rain" in the Ticino region on 13 October (personal communication from G. Kappenberger, 2002). (2) Abnormally high levels of Ca^{2+}, CO_3^{2-}, Cl^{-}, Na^{+}, K^{+}, SO_4^{2-} and Mg^{2+} in weekly bulk wet-deposition samples from the Swiss precipitation measurement stations Robiei (46°27′ N, 8°31′ E) and Piotta (46°31′ N, 8°41′ E) for the period 9–15 October (personal communication from M. Veronesi, 2002). (3) Observation of yellow snowfall on the Basodino glacier (120 km W of Piz Zupó, 46°25′ N, 8°29′ E) for the period 13–15 October 2000 (personal communication from G. Kappenberger). Therefore, the first dust event visible in the ice core most probably took place around 15 October 2000.

The dust event assigned to October 2000 was the only one in the entire core visible by its yellow colour. This event was previously detected in a shallow core drilled at the same site in May 2001. In that core the yellow layer was observed at a depth of 2.7 m w.e., resulting in an accumulation of 1.9 m w.e. during the 9-month period between the two drilling campaigns (May 2001-March 2002). This value is in very good agreement with the annual accumulation rate deduced from the entire core and thus strongly supports the dating.

The second (earlier) dust event in the ice core was at a depth of 7.1–7.3 m w.e. and had elevated levels for most chemical ions, especially MSA, Cl^-, NO_3^-, SO_4^{2-}, $C_2O_4^{2-}$, Na^+, NH_4^+, K^+, Mg^{2+}, Ca^{2+}, and H^+. This event shows remarkably high concentrations of MSA, NO_3^-, and NH_4^+, the largest within the 11-yr period covered by the ice core. δ^{18}O was relatively low during this event. Prior to this event the isotopic signature and the chemical species were low, as it is typical of winter conditions. Following the event, these indicators were characteristic of spring-summer with increased $\delta^{18}O$ values and elevated concentrations of anthropogenic species. The total suspended particle (TSP) record from nearby Jungfraujoch (not shown) was also examined and two strong episodes with TSP values $>20\,\mu g\,m^{-3}$ observed for the two-day filters 17/18 and 23/24 March, indicating that either or both of these periods may be preserved in the Piz Zupó record.

A.4 Identification of the dust transport history

In order to create a link between the specific meteorological situation leading to dust transport at Piz Zupó and the chemical signal in the ice core, we developed a refined back-trajectory method for the analysis of dust transport. It uses specific meteorological criteria to identify mobilisation and wet deposition periods of dust.

A.4.1 Back-trajectory calculations

Based on the approximate dating of the dust events described above, the potential for dust transport to Piz Zupó was studied in detail for the whole of March and October 2000. For both months, three-dimensional kinematic trajectories originating from Piz Zupó at an 6-h interval were calculated 10 days backward in time by means of the trajectory tool LAGRANTO (Wernli and Davies 1997). Together with the location of the air parcels, additional atmospheric variables, namely relative humidity, the mixing ratio of water vapour, and wind velocity were traced and stored every 6 h.

In order to account for the inherent uncertainty of trajectory calculations, the back-trajectories were calculated as ensembles arriving within a specific arrival domain. Horizontally, arrival points of trajectories were set up as a 5-point cross with one central arrival point at the location of Piz Zupó glacier, surrounded by four arrival points with an offset of 0.5° in the E, W, S, and N direction, respectively. Vertically, arrival levels ranging from 800 to 410 hPa (about 2 to 7 km a.s.l.) at intervals of 30 hPa were chosen. In total, ensembles with back-trajectories from 70 starting points were calculated daily at 00Z, 06Z, 12Z, and 18Z.

From the back-trajectories, possible areas of dust mobilisation are extracted by means of objective selection criteria. Due to the setup of the trajectory calculations, only air parcels are considered which ultimately arrive at Piz Zupó. First, air parcels are selected exhibiting conditions suitable for dust deposition at the arrival site (Sec. A.4.2), then they are traced backwards to identify the areas of potential dust mobilisation (Sec. A.4.3). The dust mobilisation locations identified from this methodology were considered as "potential" dust source regions, and verified with observational data wherever possible.

A.4.2 Dust deposition

Mineral aerosols can either be removed from air parcels by wet deposition (precipitation scavenging) or by dry deposition mechanisms. Further from the mobilisation area, wet deposition is considered to be far more effective than dry deposition (Schwikowski et al. 1995), which quickly decreases as dust is transported away from its mobilisation area (Duce et al. 1991). In fact, Osada et al. (2004) showed for a site in Japan that significant layers of mineral aerosol in snow can only form by wet deposition, typically when dusty air masses merge with precipitation-bearing frontal systems. Hence, we assume that wet deposition is most important for forming dust layers in snow at our Alpine site as well.

Consequently, in our back-trajectory analysis, we only considered those trajectories relevant for formation of a dust layer in the ice core which experienced wet deposition at the arrival site. Precipitation (and wet deposition) was assumed to occur when the relative humidity (RH) of an air parcel exceeded 80% (Fig. A.3). This threshold value follows the cloud and precipitation parameterisation scheme incorporated in the

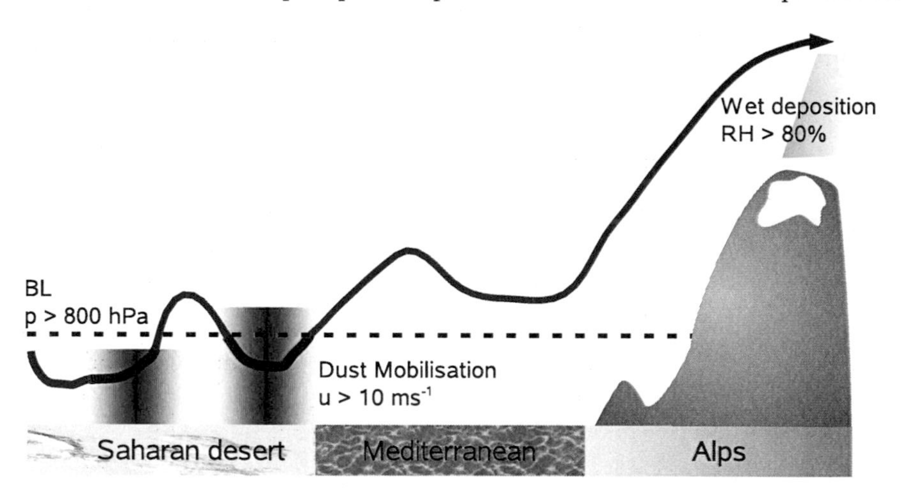

Figure A.3: Schematic of the methodology and selection criteria used for the identification of potential dust source regions from back-trajectory analysis. See text for details.

ECMWF model[3]. In addition, all air parcels of a trajectory ensemble that arrive below the uppermost "precipitating" air parcel were considered as experiencing below-cloud scavenging of mineral aerosol.

Wet removal of dust prior to arrival was estimated by counting the number of occasions in a dust-laden air parcel where RH exceeded 80% during the time period from dust mobilisation until 24 h before arrival at Piz Zupó. Independent confirmation of potential scavenging of the dust plumes along the flight path and at the arrival site was obtained from visually examining the cloud structure in Meteosat visible and IR imagery. Gravitational settling of particles along the transport path was not considered quantitatively in this study. However, in the interpretation of our results we take into account that due to this process the potential dust load is likely to decrease with increasing travel time.

A.4.3 Dust mobilisation

Again, objective criteria were adopted to extract locations along the backward trajectories where dust mobilisation potentially occurred (Fig. A.3). These are (i) the presence of dry, suspendible soil material, (ii) sparse vegetation cover, (iii) sufficiently high wind velocities near the ground, and (iv) strong updrafts which are able to lift the suspended material to a sufficient altitude for long-range transport (Shao and Leslie 1997). These criteria are also typically considered as relevant in large-scale dust transport models (Tegen and Fung 1994; Mahowald and Dufresne 2004).

According to our interpretation, criteria (i) and (ii) are fulfilled when an air parcel is located over an area in Africa north of 10° N where the vegetation map from DeFries and Townshend (1994) shows one of the categories "bare", "shrubs and bare soil", or "cultivated". Criterion (iii) is met when the air parcel is sufficiently close to the ground to pick up mineral aerosol, i.e. within the assumedly well-mixed BL. The top of the BL was defined here as the 800 hPa isosurface (~1.9 km a.s.l.). Additionally, the wind speed at the location of the air parcel within the BL has to be greater or equal to a mobilisation threshold of $10\,\mathrm{m\,s^{-1}}$. Sufficiently strong updrafts (criterion iv) are accounted for implicitly by the fact that the air parcel trajectories ultimately all arrive at Piz Zupó. Accordingly, our algorithm identifies potential dust uptake locations along a trajectory if all these criteria are fulfilled. The criteria were checked every 6 h along the individual 10-day trajectories. Thereby, a single trajectory can be associated with and hence carry dust from several potential dust uptake locations.

Sensitivity analysis showed that the dust uptake regions depend only quantitatively on the thresholds for mobilisation wind speed and boundary layer height, while the overall patterns remain the same. Currently, the aim of the proposed methodology is to provide a qualitative view of dust mobilisation and transport. Further dust-related

[3]ECMWF, 2004: IFS Documentation Cycle CY28r1, Section IV.6, http://www.ecmwf.int/research/ifsdocs/CY28r1/index.html

processes would have to be included to provide quantitative answers from these Lagrangian modelling efforts (Sect. A.7).

A.5 Meteorological development during dust transport events

Indeed, using this Lagrangian approach, dust transport events from the Sahara to the Piz Zupó glacier site could be identified during the two time periods March and October 2000, as suggested by the ice core analysis (Sec. A.3).

Detailed meteorological analyses are presented for both events which describe the specific synoptic situation associated with dust transport to central Europe. In combination with our extended back-trajectory analysis, we can reveal what types of meteorologic development were associated with dust transports to the Alps. The general transport patterns during the two dust events are illustrated by a representative selection of back-trajectories that fulfil the criteria for dust mobilisation and deposition (Fig. A.4).

Time series of variables characterising the temporal evolution of a particular Saharan dust event were derived from the enhanced back-trajectory method (Figs. A.5 and A.10). This type of visualisation combines in one figure four kinds of information: (1) the estimates of dust mobilisation in the Sahara for a particular dust event, given as the number of air parcel positions where the mobilisation criteria were fulfilled, (2) the extent of wet deposition in dust-laden air parcels prior to arrival in the Alpine area, (3)

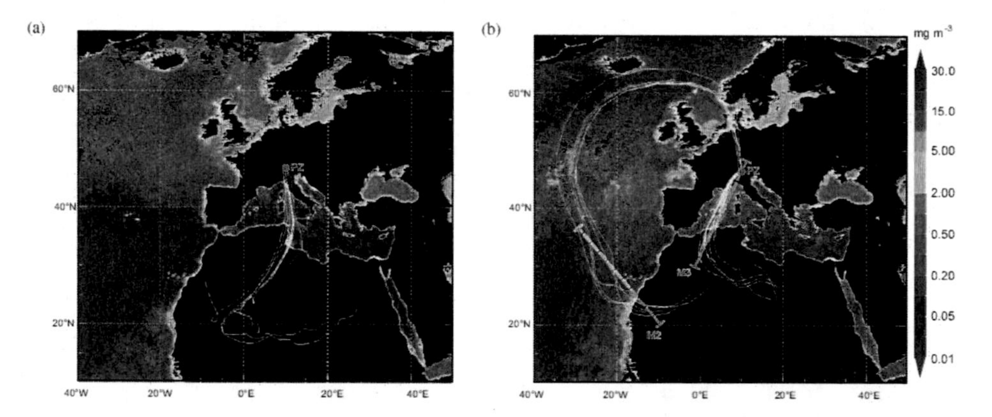

Figure A.4: Composite maps of weekly chlorophyll *a* concentration and representative examples of 10-day back-trajectories arriving at Piz Zupó (PZ) in **(a)** on 18Z 14 October and **(b)** 12Z 18 March (M2) and 06Z 24 March 2000 (M3). Cross-sections indicated in (b) by dashed, yellow lines are shown in Fig. A.14. The chlorophyll *a* maps and cross-sections are interpreted and discussed in Sect. A.6.2.

the strength of dust transport to Piz Zupó given by the number of air parcels arriving which previously experienced dust mobilisation, and (4) the precipitation at the arrival site, calculated from the decrease in specific humidity in an air parcel during the last 6 h before arrival at Piz Zupó (Wernli 1997).

Three types of visualisations are used to characterise the synoptic-scale conditions during the dust events. First, isentropic PV charts with horizontal flow vectors (only where the velocity is $>10\,\mathrm{m\,s^{-1}}$) depict the dynamical processes in the tropopause region (see for instance Fig. A.6a). The tropopause is defined as the 2-pvu surface and it separates tropospheric (blue and beige colours, low PV) from stratospheric air masses (red, yellow and green colours, high PV). Second, the state of the atmosphere in the lower troposphere is investigated with the SLP and 850 hPa equivalent potential temperature (θ_e) fields (e.g. Fig. A.6c). Third, Meteosat IR satellite images are shown for the European sector (e.g. Fig. A.6b). The exact domain for the satellite images is indicated by the white dashed frame in Figs. A.6a, c. In addition, the actual position of all trajectories that fulfilled the criteria mentioned above for potential dust transport to Piz Zupó are superimposed on the meteorological maps at the respective time instant. This permits to follow a potential dust cloud along its path from the area of mobilisation towards the deposition site concomitantly with the large-scale meteorological development. Different colours have been used to demark the trajectory position during dust mobilisation (red in the upper-level, black in the low-level charts), dust transport (yellow in the upper-level, white in the low-level charts) and wet deposition (blue). Note that dust mobilisation is only indicated for trajectories that eventually reach Piz Zupó therefore the red/black crosses do not give an overall picture of dust mobilisation.

Electronic supplements to this paper are available, which provide insightful animations of Figs. A.6 and A.11 for the full periods of the March and October dust events (see supplement zip or http://www.iac.ethz.ch/staff/harald/sahara/).

A.5.1 Meteorological development during the October dust event

During the October event (Fig. A.4a), a single dominant transport pathway leads from the Sahara across the Ligurian Sea (Fig. A.1) directly to the Alps, where it was potentially deposited at Piz Zupó between 00Z 13 October–18Z 15 October (Table A.1, O1).

Before 11 October, potential dust mobilisation was identified sporadically and dispersed across the Sahara region (Fig. A.5d). A major dust mobilisation phase occurred during 18Z 11–18Z 15 October, in a region stretching from northern Mauritania across Algeria into Tunesia (Fig. A.9a). In the eastern North Atlantic an intense low pressure system developed and moved over the British Isles at 12Z 11 October (Fig. A.6c). The satellite image (Fig. A.6b) reveals a prominent cold frontal cloud band that extends from Poland over the Alps and Spain to the Azores. At upper levels the cyclone was accompanied by a high-PV air mass on the 320 K isentrope that, at that time, induces a strong southwesterly flow to the Alps (Fig. A.6a). Further south, strengthening low-

Table A.1: Potential timing and source regions of the Saharan dust mobilisation and deposition phases for the March and October 2000 dust events as identified from the extended back-trajectory analysis and visual inspection of IR satellite imagery.

Phase	Mobilisation	Deposition	Source regions
M1	00Z 06–18Z 07 Mar 2000	06Z 14–18Z 15 Mar 2000	Mauritania/Algeria
M2	18Z 08–06Z 12 Mar 2000	06Z 16–12Z 19 Mar 2000	Algeria/Mauritania/Mali
M3	18Z 19–00Z 23 Mar 2000	12Z 23–12Z 26 Mar 2000	Algeria/Libya
O1	18Z 11–18Z 15 Oct 2000	00Z 13–18Z 15 Oct 2000	Algeria/Libya

level winds across the western Sahara led to dust mobilisation with some dust already entering the Mediterranean. During the next two days, a southwest-to-northeast oriented mid-level jet (with peak velocities of more than 30 m s^{-1} between 500 and 800 hPa, not shown) developed and led to an intensification of dust mobilisation over Algeria during 12–13 October (Fig. A.5d).

Until 12Z 13 October the upper-level PV anomaly elongated meridionally and developed into a so-called PV streamer that extended to Northern Africa (Fig. A.6d). Along its eastern side, the streamer veered the flow from Algeria directly towards the Alps. Near the surface, a tongue with low θ_e was advected from the North towards Spain and Morocco, and a tongue with high θ_e from Africa over the Alps to the Baltic Sea. A sharp cold front, running parallel to the upper-level streamer demarked the separation between these strongly differing air masses (Fig. A.6f). It was associated with a relatively broad cloud band extending from Algeria over the Alps to northern Germany (Fig. A.6e). At that time, there had been heavy precipitation for more than 24 h along the Alpine southside (Fig. A.7). This heavy precipitation period is also apparent in the precipitation extracted from our back-trajectory analysis (Fig. A.5a). Ahead of the upper-level streamer, a large dust plume had entered the central Mediterranean and some wet deposition already occurred in the Alps. The SeaWiFS visible image from 13 October (Fig. A.8) indicates the presence of adjacent streaks of dry dusty and moist cloudy air masses that may have started mixing on their way to the Alps. The trajectory analysis indicates that at Piz Zupó the dust event sets in rather abruptly after 00Z 13 October (Fig. A.5c). At that time, dust mobilisation continued near Tunesia. The very short time span between dust mobilisation and deposition of about 24–48 h reflects the rapid northward transport during this particular meteorological situation.

During the next two days, until 12Z 15 October, the PV streamer broke up and evolved into a PV cut-off located over the western Mediterranean (Fig. A.6g). It did not reach into northern Africa and the export of Saharan dust into the Mediterranean ceased (Fig. A.5d). However, previously mobilised dust was transported on its eastern flank from S-SE directions towards the Alps where widespread wet deposition occurred (Fig. A.5b). The surface front deformed strongly (Fig. A.6i) and a large-scale cloud system with embedded convection covered all of Italy and France (Fig. A.6h). It appears

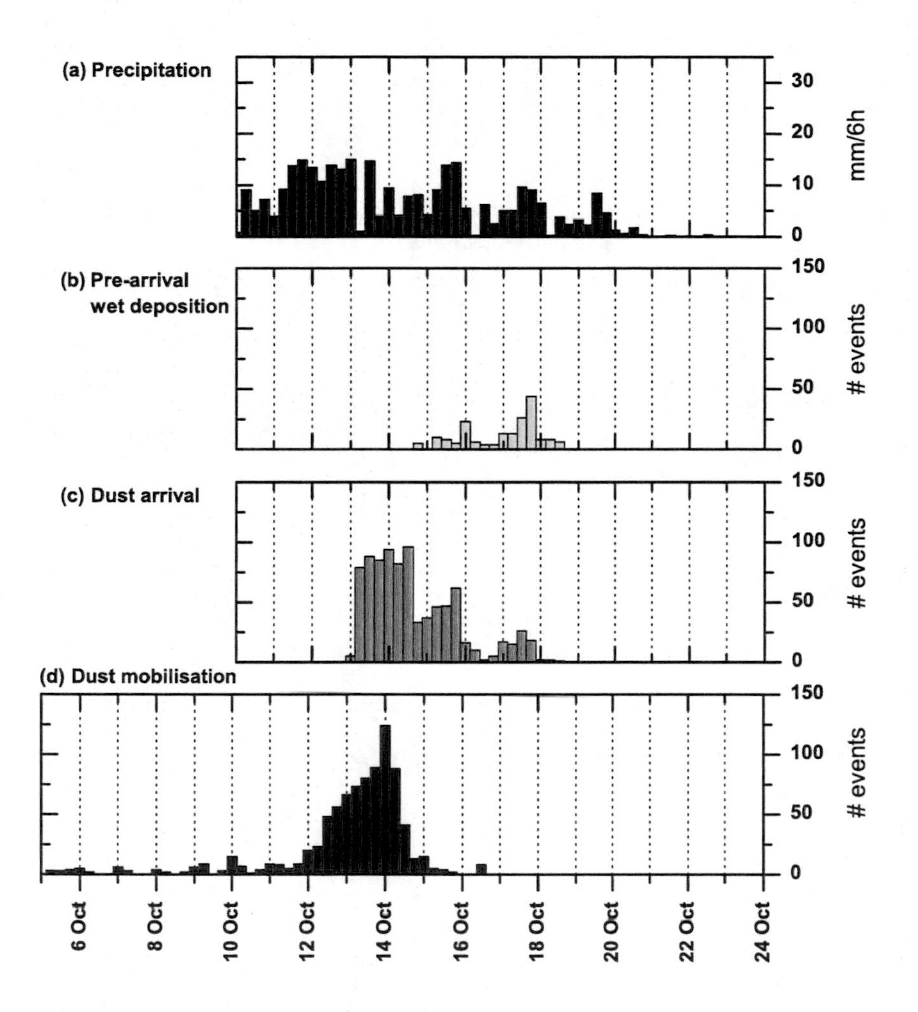

Figure A.5: Temporal evolution of dust transport characteristics extracted from the extended back-trajectory analysis for the October event. **(a)** precipitation estimate, **(b)** wet deposition until 24 h before arrival, **(c)** arriving dust-laden air parcels, **(d)** dust mobilisations.

to be responsible for the more intermittent rainfall observed near Piz Zupó between 13 and 17 October (Fig. A.7), possibly reflecting the arrival of mixed dusty and moist air masses. A more intermittent precipitation regime is also apparent in the trajectory-based precipitation series (Fig. A.5a). By 18Z 15 October, the last of the identified dust parcels have arrived at Piz Zupó (Fig. A.5c). During the same period, wet deposition prior to arrival became increasingly wide-spread (Fig. A.5b), and thereafter little precipitation was observed at the three stations until the end of the month.

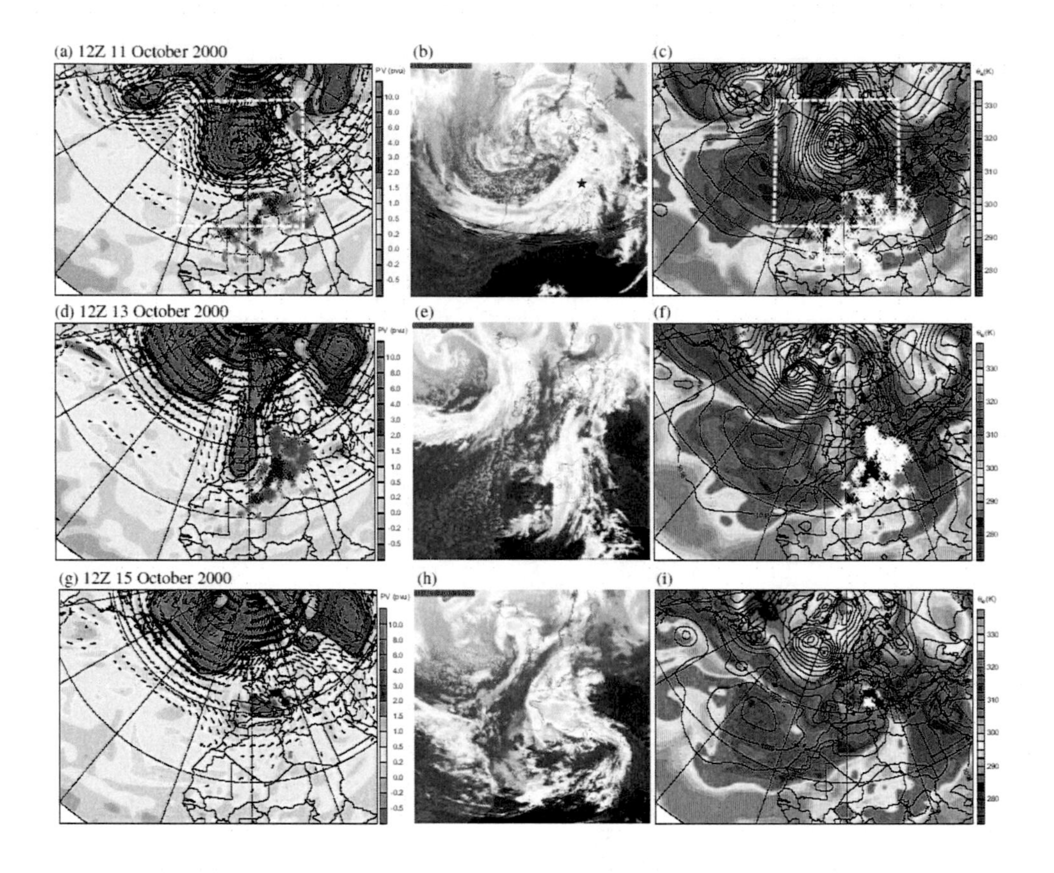

Figure A.6: Meteorological situation during the October 2000 dust event (11 to 15 October 2000). Left to right column: Potential vorticity (PV) and wind vectors at 320 K, Meteosat IR satellite image, θ_e and sea level pressure (SLP) at 850 hPa. Red star in (a-c) shows the approximate location of Piz Zupó.For details see text. See electronic supplement to this paper for a movie of this Figure.

A.5.2 Meteorological development during the March dust event

For the more complex March event trajectory analysis indicates a first minor dust phase (M1) (not shown) followed by two important episodes (M2, M3) characterised by different transport pathways (see Table A.1 and Fig. A.4b). During M2, dust was ejected across the African west coast, transported in a rather exceptional pathway over the eastern North Atlantic, and reached about 60° N before approaching the arrival site from northerly directions. In contrast, transport during M3 was again mainly northward, similar to the October event.

The first minor dust mobilisation phase in March 2000 (M1, cf. Table A.1) took place in response to a cyclonic system located over the Canary Islands. It first mobilised dust in the border region between Algeria, Mali and Mauritania and led to the ejec-

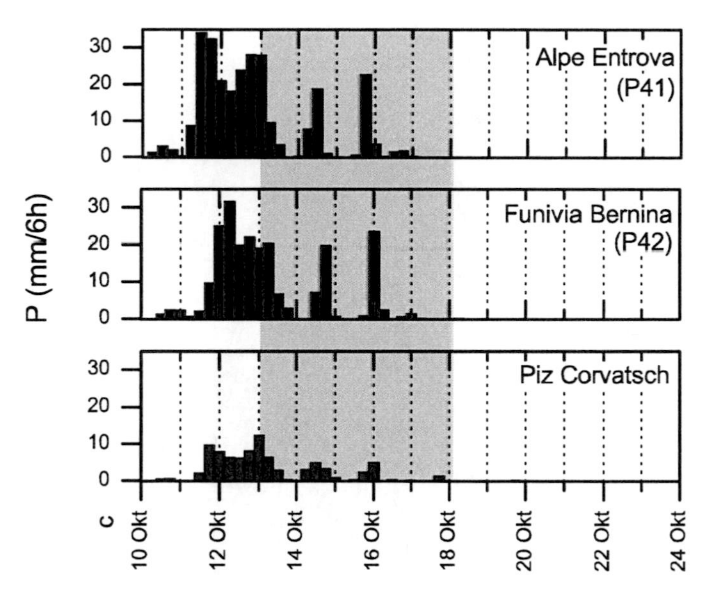

Figure A.7: Six-hourly accumulated precipitation at three meteorological stations south (P41, P42) and north (Piz Corvatsch) of Piz Zupó for 10–18 October 2000. The gray shaded area indicates the dust arrival period identified from the back-trajectory analysis (Fig. A.5c).

tion of a dust plume into the Atlantic during 12Z 05–18Z 07 March (Fig. A.10d). The cyclone is still visible on the satellite image from 12Z 11 March over southern Portugal (Fig. A.11b) associated with a small-scale upper-level PV cut-off (Fig. A.11a) and a weak signal in the SLP field (Fig. A.11c). The ejected dust parcels were picked up by the westerlies to the west of Ireland after 11 March and were advected across France towards the Alps. This principal transport pattern was already described by Prodi and Fea (1978), and Goudie and Middleton (2001) list numerous Saharan dust falls over the British Isles during the twentieth century, including an event in Oxfordshire on 13 March 2000 which was probably related to the M1 dust plume. In a Lagrangian dust transport study, Ryall et al. (2002) also modelled Saharan dust transport to Great Britain on, among other days, 13–14 March 2000. However, widespread rain occurred among the dust-laden air parcels during the last 72 h before arriving at the Alps. This is reflected by the pre-arrival wet deposition detected for the M1 event (Fig. A.10b). It is therefore unlikely that a significant amount of dust from phase M1 was deposited at Piz Zupó during 06Z 14–18Z 15 March.

A second major dust mobilisation phase (M2) was identified from the trajectory analysis during 18Z 08–06Z 12 March (Fig. A.10d), first in south-western Algeria and Mali, later also in Mauritania (Figs. A.11a, A.9b). During 12Z 10–06Z 12 March a large dust plume left Africa near the Capverde Islands (20° N, 20° W). This is confirmed by the SeaWiFS satellite imagery (Fig. A.12a) and by the 3rd-level product from SeaWiFS, the Aerosol Optical Thickness one-week composite (not shown). On 14 March the dust plume was picked up by a weak cyclone in the eastern North Atlantic (near 40° N,

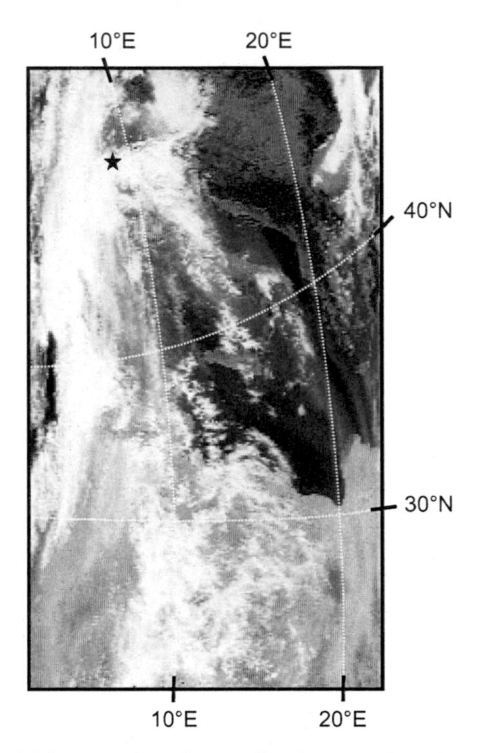

Figure A.8: SeaWiFS visible wavelength satellite image of a dust plume over the central Mediterranean on 13 October 2000. Red star shows the approximate location of Piz Zupó.

40° E, see Figs. A.11d, f). The decaying cyclone was associated with a minimum in SLP, a uniform low-level temperature field and a small upper-level PV cut-off. This feature was essential to advect the Saharan dust northward into a pronounced upper-level ridge and jet-stream system. Almost no clouds and rain formed in the vicinity of the dust plume until it passed Iceland to the north of a pronounced surface high-pressure system on 12Z 16 March (Figs. A.11g, i). At that time a strong northerly flow was established towards the Alps, as can be seen in the IR satellite image by the sharp transition between orographic clouds and cloud-free conditions south of the Alps and Pyrenees due to Föhn (Fig. A.11h). During the following two days, this northerly flow advected the dust cloud rapidly towards Central Europe. Partial rain-out occurred over Germany (see Fig. A.10b and electronic supplement). About 9 days after mobilisation, the remainder of dust plume M2 arrived at Piz Zupó during 12Z 17–06Z 19 March (Fig. A.10c), considerably later than for the October event. Precipitation data close to Piz Zupó indicate that wet deposition of the remaining dust may have occurred during 17–18 March, although precipitation intensities were small and limited to high elevations and the slopes facing north (Fig. A.13). The trajectory-based precipitation estimate also shows small precipitation intensities during this period (Fig. A.10a).

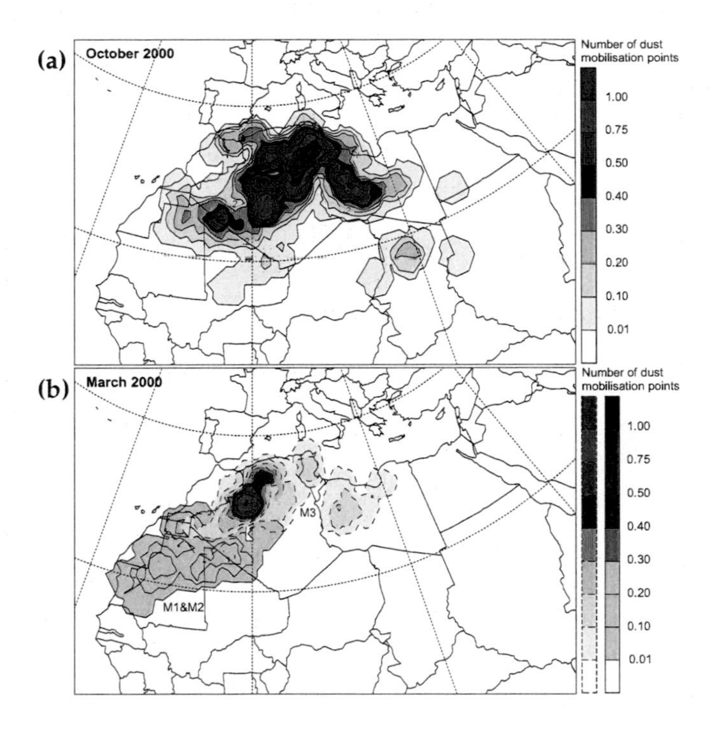

Figure A.9: Probability density functions of potential source regions of Saharan dust identified from objective selection criteria for the Saharan dust events recorded in the Piz Zupó ice core in (a) October and (b) March 2000. Scale in arbitrary units. See text for details.

A third pronounced dust mobilisation phase (M3) between 18Z 19–00Z 23 March (Fig. A.10d) was confined to a narrow band located across the Algerian desert (Figs. A.11j, A.9b). Similarly to the October event, the dust mobilisation appears to be closely related to a PV streamer reaching into northern Africa, which induced strong southerly winds on its eastern flank. As an interesting side remark it is noted that this PV streamer developed in a very different way compared to the streamer in the October event. The latter evolved due to narrowing of an upper-level trough. In contrast, in March a previous large streamer extending from the Black Sea over the entire Mediterranean broke up on 21 March near the Atlas mountains. The cut-off PV vortex rotated and moved slightly northward until it rejoined the stratospheric reservoir near Scotland at 00Z 22 March (Fig. A.11j). This "secondary streamer" led to the ejection of the dust plume into the Mediterranean during 18Z 21–12Z 23 March. Its local PV maximum coincided with a weak surface cyclone west of Portugal (Fig. A.11l) that can also be identified in the IR image with a broad and fuzzy cloud structure in the warm sector (Fig. A.11k). The dust ejection into the Mediterranean is confirmed by the visible satellite image on 22 March (Fig. A.12b). During 12Z 23–00Z 25 March wet deposition took place at Piz Zupó in qualitative agreement with the cloud structures in the IR images (see electronic supplement), the scattered precipitation signals recorded in the Piz

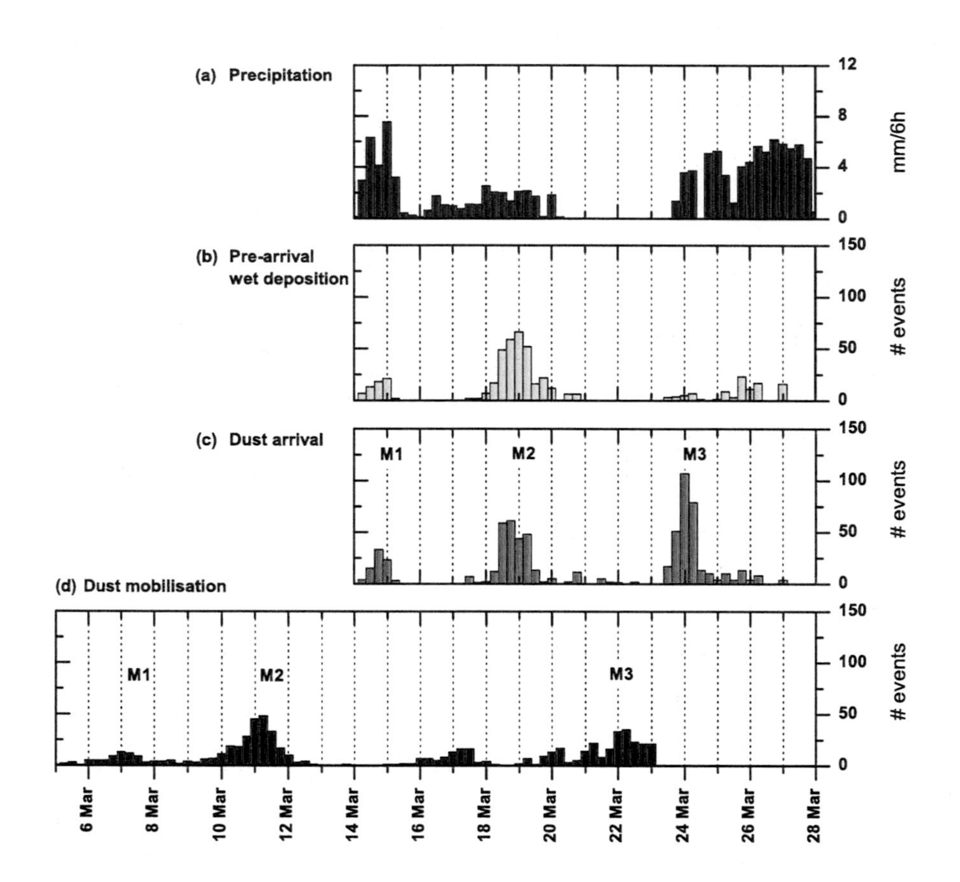

Figure A.10: Temporal evolution of dust transport characteristics extracted from the extended back-trajectory analysis for the March event. **(a)** precipitation estimate, **(b)** wet deposition until 24 h before arrival, **(c)** arriving dust-laden air parcels, **(d)** dust mobilisation.

Zupó area during this time period (Fig. A.13), and the trajectory-based precipitation estimate (Fig. A.10a). Wet deposition prior to arrival is low for the strongest part of the M3 phase, indicating that conditions were favourable for preserving this dust event at Piz Zupó.

A.5.3 Large-scale flow imprints on dust transport

The above analysis of the meteorological flow evolution during the October and March dust events demonstrates that there is no archetypal pattern which leads to dust events in the Alpine area. The enhanced trajectory analysis allowed for distinguishing dust events which were likely to be preserved in the glacier from those that probably did not

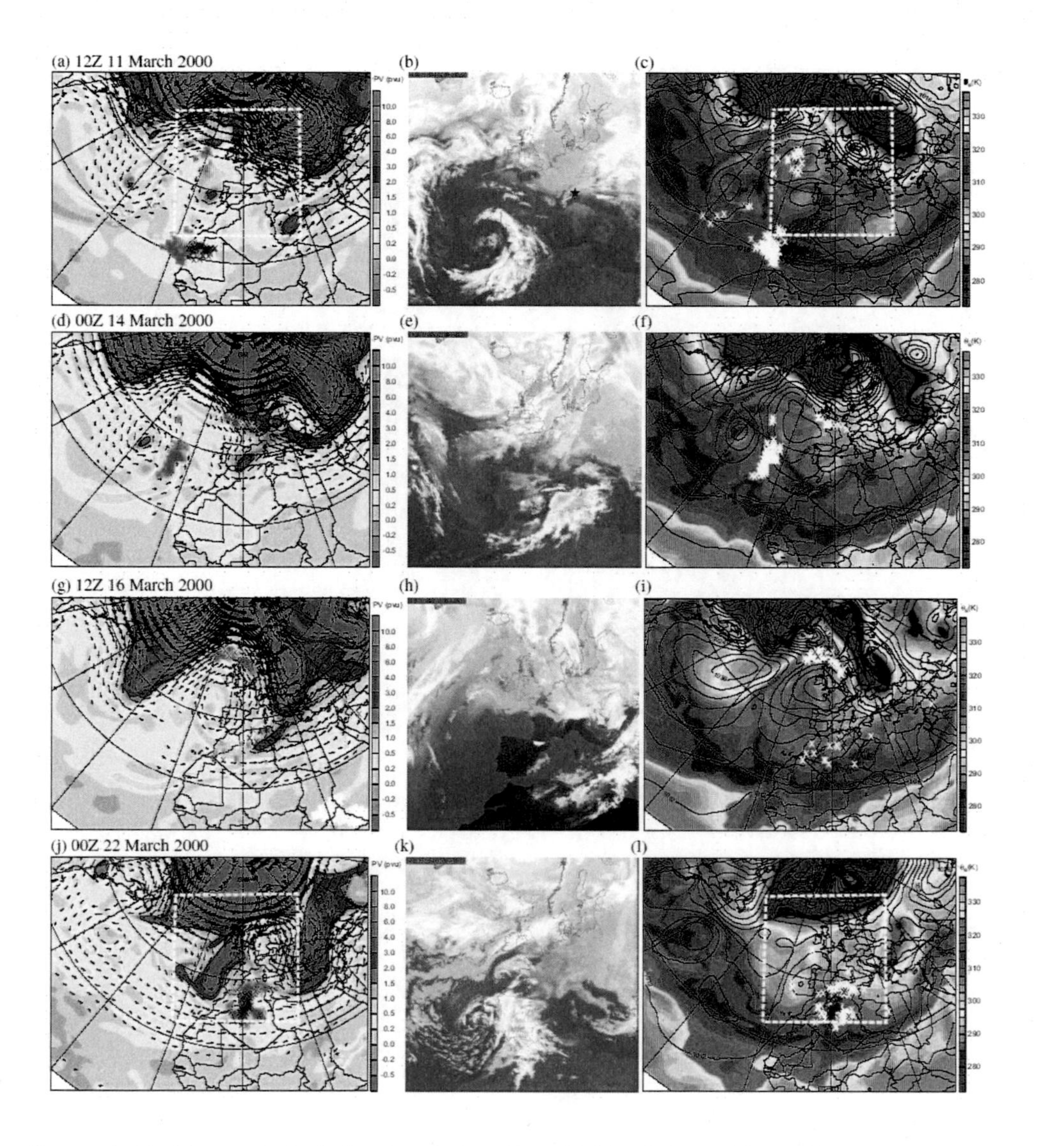

Figure A.11: Meteorological situation during 11 to 16 March 2000 (Phase M2) and 22 March 2000 (Phase M3). Left to right: PV and wind vectors at 315 K, infrared satellite image, θ_e and sea level pressure (SLP) at 850 hPa. Red star in (a-c) shows the approximate location of Piz Zupó.For details see text. See electronic supplement to this paper for a movie of this Figure.

reach Piz Zupó due to prior wet deposition, and to attribute a specific flow evolution to each dust transport event.

Remarkable similarities and differences between the considered phases of dust export are apparent. First, direct northward transport from the Sahara to the Alps oc-

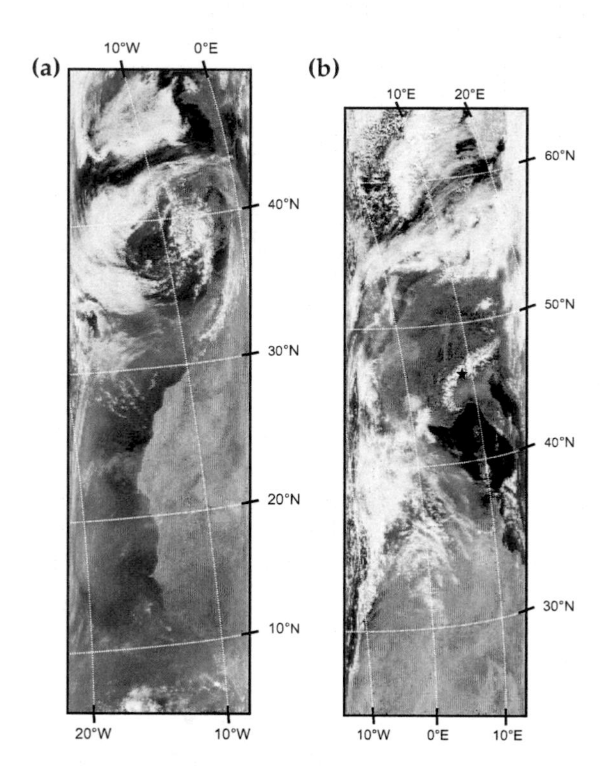

Figure A.12: SeaWiFS visible wavelength satellite images of **(a)** the dust plume of phase M2 on 12 March 2000 to the west of Africa, **(b)** the dust plume of phase M3 on 22 March 2000 in the Mediterranean. Red star shows the approximate location of Piz Zupó.

curred when a stratospheric PV streamer was stretching into northern Africa and induced anomalous southerly flow over the Mediterranean. In the October case, this streamer formed due to Rossby-wave breaking (see an example in Morgenstern and Davies 1999), while in March during phase M3 a first streamer of type I (Appenzeller et al. 1996) broke up and formed a cut-off that remerged near Scotland to form a secondary streamer. This difference in the formation history of the two western Mediterranean PV streamers is noteworthy and illustrates the complexity and variability of meteorological processes that can lead to large-scale dust transport to Central Europe. This aspect is underlined by the dust transport phase M2 in March 2000, where several seemingly independent meteorological features cooperated in order to transport dust from the western Sahara within 9 days across the Capverde Islands and the Azores to 60° N and across Germany to Piz Zupó.

The importance of the upper-level flow configuration on dust export from the Sahara has already been noted by a number of studies (Prodi and Fea 1978, 1979; Alpert and Ganor 1993; Barkan et al. 2005). Notably, the PV streamer flow configuration is reminiscent of the typical precursor structure for heavy precipitation on the Alpine

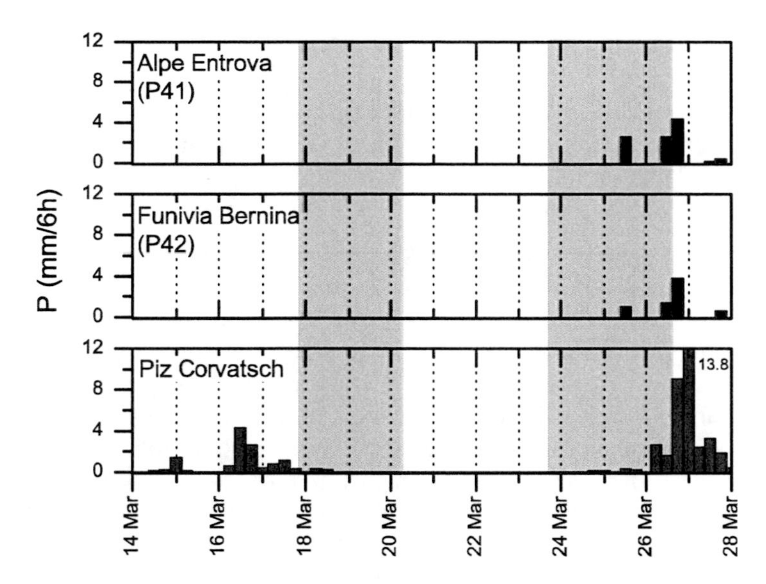

Figure A.13: As Fig. A.7, but for 14–28 March 2000. Areas shaded in gray from left to right denote the dust phases M2 and M3 identified from the back-trajectory analysis.

south-side, where quasi-stationary PV streamers extending over France into the western Mediterranean can direct large moisture fluxes along their eastern side towards the Alps (Massacand et al. 1998; Martius et al. 2006). From our study, it becomes evident that particularly elongated streamers which also reach into northern Africa can direct dry and dusty Saharan air masses towards the Alps. Similar atmospheric flow configurations including an upper-level PV streamer extending into the African subtropics and a southwest-northeast oriented jet in the middle troposphere may also play a role during tropical-extratropical interactions (Knippertz et al. 2003). Convective activity triggered near the right entry and left exit of the jet could further support uplift of desert dust and subsequent long-range transport in adjacent bands of moist cloudy and dry dusty air towards northerly directions (Figs. A.8, A.6e). In some cases, the northward transport of dust and moisture occurs almost simultaneously, as indicated by the severe flash flood event in Gondo on 13/14 October 2000 that was followed by the most prominent dust event recorded in the Piz Zupó ice core. Note however that this combined occurrence of heavy rain and Saharan dust in the southern Alps might be exceptional since (i) many PV streamers that produce heavy rain do not reach far enough into the desert to trigger the export of dust, (ii) not every southerly flow across the Mediterranean that carries Saharan dust is also advecting large amounts of moisture from the Mediterranean. It is therefore not possible to interpret the ice core's dust signal as a direct record of extreme precipitation, or vice versa.

A.6 Contributions to the chemical signal in the ice core

The chemical signal in the ice core potentially reflects contributions from the source area of dust, from different transport paths, and from differences in the deposition process. Applying again the extended back-trajectory methodology, we aim to disentangle the importance of the various contributors to the ice core chemistry.

An important source of information for this task are additional chemistry data at higher temporal resolution than provided by the ice core, for instance from aerosol or precipitation chemistry measurements. No chemical analyses of the precipitation near Piz Zupó are available for comparison. Instead, we rely on data from the literature which were acquired at other Alpine locations as a proxy for Piz Zupó. During February and March 2000, the Cloud and Aerosol Characterisation Experiment (CLACE-1) took place at the Jungfraujoch (JFJ) high alpine observatory (46°32′N, 7°59′E, 3580 m a.s.l., Fig. A.1). During this campaign, several physical and chemical properties of aerosol were sampled during both dust phases M2 and M3 at very high temporal resolution. Besides scattering and absorption coefficient measurements, single-particle composition spectra were acquired (Hinz et al. 2005).

Back-trajectory analysis repeated for the JFJ revealed some differences in timing and magnitude of the dust events compared with Piz Zupó (not shown). At JFJ, the dust phases occurred slightly earlier than at Piz Zupó (Phase M2: 20Z 16–20Z 18 March from CLACE-1, 12Z 17–18Z 19 March from JFJ back-trajectories, phase M3: 03Z 23–18Z 24 March from CLACE-1, 06Z 23–00Z 25 March from JFJ back-trajectories). Dust phase M2 experienced less pre-arrival wet deposition, and the M3 event was significantly stronger than at Piz Zupó. The CLACE-1 measurements thus provide some independent evidence for the chemical composition of the arriving dust and the results from our trajectory study. However, small-scale variability inherent to Saharan dust events in the Alpine area makes a detailed comparison between JFJ and Piz Zupó difficult. Therefore, we can only use the CLACE-1 data as a proxy information for the actual precipitation composition at Piz Zupó.

A.6.1 Dust source region signal

The enhanced back-trajectory method allowed for extraction of the locations where the mobilisation criteria as defined in Sect. A.4.3 were fulfilled. Probability density functions of these potential uptake locations are interpreted as potential source areas of Saharan dust (Fig. A.9). Uptake locations are only shown for air masses transported towards Piz Zupó, and for the mobilisation periods given in Table A.1.

For the October 2000 event (Fig. A.9a), the identified source regions are rather widely dispersed in the northern Sahara. Yet, two regions with increased potential mobilisation are apparent, one in Algeria along the axis of the previously mentioned mid-level jet, and one in western Libya. These northern mobilisation maxima together with the wider distribution of weaker sources indicate that although the mineralogy of soils in the northern Sahara dominates, dust mobilised previously further south may be mixed into the dust plume transported to the Alps. This could lead to a blurred chemical source region signal (Schütz and Sebert 1987).

For the March 2000 events, dust source regions were identified separately for the dust phases M1, M2 and M3 (Fig. A.9b). The first two phases exhibit dust sources in West Sahara, Mali, and Mauritania, while for the phase M3 source regions in Algeria and Libya dominate, which is indicative of the meteorological similarity to the October event.

In the Piz Zupó ice core record, large peaks in the concentration of Ca^{2+} mark the October and March dust events (Fig. A.2). The Ca^{2+} concentration during the October event is by far the largest signal during the 11-yr record. It covers about 0.20 m w.e. in the ice core. The section identified as the March event in the ice core shows two peaks in Ca^{2+} close to each other, a stronger one from 7.1–7.3 m w.e. and a smaller one from 7.0–7.1 m w.e. Soils in the northern Sahara are relatively rich (partly $>15\%$) in calcite ($CaCO_3$), while towards the south concentrations decrease (Schütz and Sebert 1987; Claquin et al. 1999). Accordingly, Ca^{2+} and calcite are widely used as tracers of Saharan dust events over Europe (Wagenbach and Geis 1989; Schwikowski et al. 1995; Avila et al. 1997), and the high concentrations for the October and March events are in good agreement with the prominent northern Saharan dust source regions identified by the enhanced back-trajectory analysis.

Single-particle mass spectra measured during CLACE-1 support the shift in source regions between phase M1/M2 and M3. Hinz et al. (2005) found differences in the intensity ratios of nitrate, phosphate, sulfate, silicon, titanium, and iron and their oxides between 17/18 and 23 March. In particular, slightly higher class abundances of particles rich in iron oxides (particle class 'mineral 2', Fig. 8 in Hinz et al. (2005)) were noted on March 23.

However, due to resuspension, the mineralogy of a mineral aerosol is not necessarily identical with the mineralogy of its source region (Schütz and Sebert 1987; Claquin et al. 1999). Furthermore, changes in the mineralogical composition might occur during dust transport due to the preferential deposition of coarser particles or dissolution of calcite. This could in particular be the case for long transport paths, such as for phase M2. The analysis of rare earth elements, and the $^{87}Sr/^{86}Sr$ and $^{143}Nd/^{144}Nd$ combined isotope ratios could improve the identification of source regions, but so far very few such data are available from Alpine ice cores.

A.6.2 Uptake of chemical species during transport

Several chemical species which show extraordinarily high values in the Piz Zupó ice core for the two dust events (Sect. A.3, Fig. A.2) are unlikely to originate from the source region of the mineral aerosol. This includes typical sea spray components, such as Na^+, Cl^-, and partly SO_4^{2-}, as well as MSA. In a laboratory study, Adams et al. (2005) showed that Saharan dust can effectively bind SO_2, which later can be oxidised into SO_4^{2-}. MSA is formed by photo-oxidation of dimethylsulfide (DMS), which is in turn released by marine phytoplankton near the ocean surface (Huebert et al. 2004). The chemical species NH_4^+, NO_3^-, and partly SO_4^{2-} are tracers for polluted air masses (Schwikowski et al. 1995; Jordan et al. 2003; Aymoz et al. 2004). Their strongest sources are in the marine or anthropogenically influenced BL.

Accordingly, these chemical components are most likely scavenged by the mineral aerosol somewhere along its transport path. Scavenging can take place in several ways: (i) Chemical components are contained in BL air masses that are transported to the glacier independently from the dust transport, where they then undergo scavenging during wet deposition. Such independent yet subsequent transport and deposition of dust and anthropogenic pollution has recently been reported recently for a valley in the French Alps (Aymoz et al. 2004). (ii) Chemical components are scavenged by the mineral aerosol particles along their path through the BL, and subsequently deposited at the glacier adsorbed onto the mineral aerosol. (iii) Chemical components are already contained in the air mass before dust mobilisation occurs, and scavenging takes place during mobilisation and transport. A combination of these processes may be possible as well.

Sea salt

With respect to sea salt species, the ice core record shows only minor differences for the two dust events (Fig. A.2), the exception being Cl^-, which is much larger for the October than for the March event. The high concentrations of NO_3^- and NH_4^+ during March are also exceptional. Schwikowski et al. (1995) noted extraordinarily high scavenging efficiencies of Cl^-, NO_3^-, SO_4^{2-}, and NH_4^+ during snow falls associated with Saharan dust at JFJ, probably due to increased riming. For TSP data from JFJ, Henning et al. (2003) note a general enhancement in the non-volatile nitrate fraction with increasing calcium concentrations due to reaction of gaseous HNO_3 with $CaCO_3$, but also some situations where small coarse nitrate fractions occur along with high Ca^{2+}, namely during Saharan dust events. Further comparison of specific concentration measurements at JFJ with our ice core data is prohibited by the pronounced spatial variability inherent to Saharan dust events on scales of $\approx$10 to 100 km. Nevertheless, it seems plausible that during both dust events contact or mixing with marine BL air occurred, while during the March event in addition polluted air masses interacted with the dust plume somewhere along the transport path.

MSA

A different explanation is required for the MSA signal in the ice core. While for the March event a large MSA peak similar to that of the anthropogenic pollutants can be observed, MSA is virtually absent from the ice core record during October. The question arises as to what causes the difference in MSA between the two dust events. This issue is addressed by considering the potential source regions of MSA during the two dust events. Since MSA resides in the marine BL as an oxidation product of DMS, which is exclusively produced by certain phytoplankton species, we use chlorophyll *a* concentrations at the sea surface obtained from remote sensing satellites as a proxy for the presence of MSA in the marine BL (Sect. A.2.3).

Overlaying the chlo-*a* maps with representative dust transport trajectories for the two dust events indicates potential for MSA uptake from areas of high phytoplankton concentrations (Fig. A.4). During the week 7–14 October, areas of high concentration of chlo-*a* are confined to the North Sea and upwelling areas off the African west coast (Fig. A.4a). In these areas, concentrations of chlo-*a* reach 3–10 mg m^{-3}. A composite for the week from 13–20 March (Fig. A.4b) shows a general southward shift of high phytoplankton concentration areas, and an intensified maximum off the African west coast. Importantly, the western Mediterranean now exhibits a centre of moderate concentration of chlo-*a* between 0.3–1.0 mg m^{-3}, compared to less than 0.1 mg m^{-3} in October. This is an indication of the spring phytoplankton bloom that usually takes place in this area during March and April (Bosc et al. 2004). Most other areas of the Mediterranean show low concentrations of chlo-*a* at that time of the year (<0.2 mg m^{-3}).

Time-series of vertical cross-sections as indicated in Fig. A.4b were inspected for the passage of the dust plumes across the potential source regions of MSA, and in order to identify the likelihood of an interaction between the dust plume and the potentially MSA-laden marine BL air masses (Fig. A.14). During the dust phase M2, the extended back-trajectory analysis indicates that the dust-laden air masses offshore the African

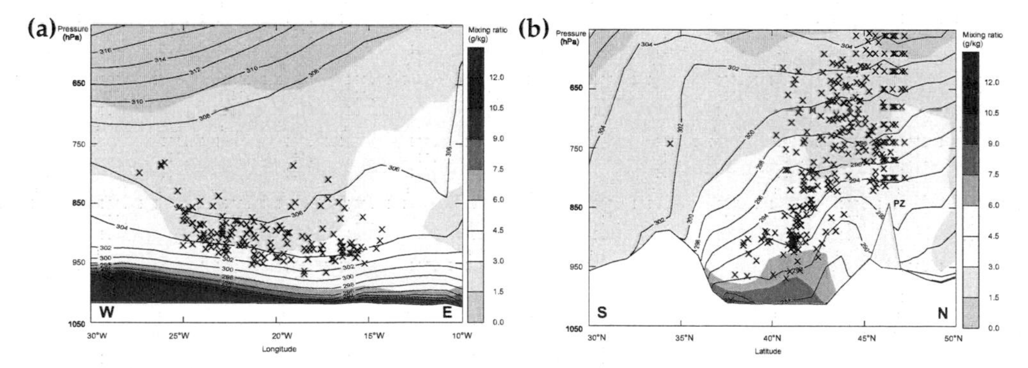

Figure A.14: Vertical cross-sections during the ejection of the dust plumes along the transects marked in Fig. A.4b. **(a)** M2 (00Z 12 March) and **(b)** M3 (18Z 23 March). Shown are mixing ratio (g kg^{-1}, shaded), potential temperature (contour interval 2 K, contours), and crosses indicating the position of potentially dust-laden air parcels identified from the back-trajectory analysis. The vertical axis is given in pressure units; 800 hPa correspond roughly to 1900 m a.s.l. and 500 hPa to 5500 m a.s.l. The marine boundary layer top is expected near the strongest vertical gradient in mixing ratio.

west coast were above the marine BL (Fig. A.14a). The marine BL top is defined here by the strongest vertical gradient of water vapour mixing ratio. The dry Saharan air masses continuously remained above the very shallow marine BL and subsequently were slowly lifted during their northward transport (Sect. A.5.2). While crossing the potential source areas of MSA in the North Sea, the dust plume was still above the marine BL (850–600 hPa, not shown). However, during the dust phase M3, a different picture emerges (Fig. A.14b). Here, the marine BL over the western Mediterranean is less clearly defined than off the West African coast. Some dust parcels appear to travel within moist areas of the marine BL. The air masses approach Piz Zupó directly across the chlo-*a* maximum in the Gulf of Genova (Fig. A.4b). Rapid dust ejection governed by a mid-tropospheric jet structure similar to the October event possibly triggered convection near the jet's left exit region during 23 March (see electronic supplement), thereby enhancing the likelyhood of mixing between mineral aerosol and marine BL air. Hence, our back-trajectory analysis indicates that the large MSA peak in the ice core record for the March event was caused by aerosol scavenging while the dust plume M3 interacted with the marine BL air during its passage over the spring maximum in the phytoplankton bloom in the western Mediterranean.

Single particle analysis during the CLACE-1 campaign at JFJ showed an interesting shift in chemical composition from 23 to 24 March (Hinz et al. 2005, Figs. 8,10). On 23 March, particles reflecting an internal mixture of carbon and mineral components were found, albeit in low concentrations. These indicate interaction of carbon-containing particles with mineral particles. Towards 24 March, incidence abundances of ammonium, sulfate, nitrate, and carbon increased while mineral components showed a decrease. These findings from Hinz et al. (2005) corroborate that phase M3 brought chemically altered mineral aerosol particles, which had been exposed to BL air both over the Mediterranean and northern Italy into the Alpine area. Back-trajectory analysis confirmed that the transport of Saharan dust at JFJ quickly ceased after 06Z 24 Mar (not shown). The transport path of dust-laden air parcels suggests that pollution sources near Genova or Milano contributed to the anthropogenic signals at Piz Zupó.

Vertical cross-sections for the October event (not shown) show similar potential for interaction with the marine BL over the Mediterranean as during March. High concentrations of sea spray components (e.g. Cl^-) in the ice core chemistry confirm that such interaction occurred (Fig. A.2). However, as can be inferred from Fig. A.4a, in October concentrations of chlo-*a* were substantially lower in the western Mediterranean than during March. This would also be expected from the phytoplankton seasonality in that area (Bosc et al. 2004). Therefore, the most likely explanation for the absence of MSA in October is that concentrations of MSA in the marine BL were much smaller than in March.

A.7 Discussion of the applied methodology

An assessment of the reliability and soundness of the new methodology applied here is required in order to draw significant conclusions from this study, in particular as it combines data from different sources and with different spatial and temporal scales. The ice core from Piz Zupó represents a point sample that integrates over time scales of a few days. Back-trajectory calculations, on the other hand, while having a higher temporal resolution, are limited by their spatial accuracy and representativeness, and are not straightforwardly related to a point sample. Some critical aspects related to the methodology are elucidated below.

The spatial uncertainty due to uncertainties in the wind fields and calculation techniques is a central issue of back-trajectory methods. Past dust transport studies mostly relied on single back-trajectories, typically at different vertical levels (Schwikowski et al. 1995; Avila et al. 1997; Ansmann et al. 2003). In this study, spatial uncertainty is taken into account by calculating ensembles of trajectories for each time step (Sect. A.4.1), rather than considering single starting points. However, spatial uncertainty increases with the integration time of backward trajectories. The spatial spread of the ensemble gives an indication of the coherence of the flow and hence the reliability of e.g. a potential dust source region. The 10-day calculation time is therefore a compromise between the reliability of the calculations and the ability of the method to capture all relevant dust mobilisations. Note, however, that this method cannot account for analysis errors in the wind field, e.g. due to sparse data coverage.

Sufficiently high wind velocity near the surface is a well-established criterion for dust mobilisation. Studies by Shao and Leslie (1997) and Schoenfeldt and von Loewis (2003) indicate that if the 10 m wind speed exceeds $10\,\mathrm{m\,s^{-1}}$, wind shear near the surface becomes sufficiently strong to mobilise significant amounts of dust. The wind speed considered by our methodology is taken at the position of an air parcel rather than at a fixed height of 10 m above ground, hence our mobilisation velocities may be biased high. However, observing the movement of Saharan dust over land and sea from satellites, Koren and Kaufman (2004) found velocities of 10 to $13\,\mathrm{m\,s^{-1}}$ for whole dust plumes.

Our assumption for the boundary layer height (BLH) is well in the range typically observed and modelled for the Sahara. In a sensitivity study of the TOMS aerosol index, Mahowald and Dufresne (2004) found boundary layer heights of around 1 km in January and larger than 3 km in July. As we consider the months March and October only, a BLH of $\approx$1.9 km a.s.l. is a reasonable assumption. However, we do not account for the pronounced daily cycle in BLH in desert areas, with nighttime values as low as 200 m (Mahowald and Dufresne 2004). During the strong wind conditions required for the mobilisation of dust, however, nighttime BLHs should be larger than the low BLHs produced by radiative cooling in steady air. In a number of LIDAR studies, the height of dust plumes leaving the African continent was typically observed to be at or above 1.5 km (Hamonou et al. 1999; Mattis et al. 2002; Leon et al. 2003). Off the African coast, dust layers were in general multi-layered, and reached up to 5 km above ground. Hence, long-range transport seems to begin at around 1.5 km height above ground. Taking into account the underlying orography, our assumed BLH complies well with these observations.

Soil moisture is a factor that can be important for dust mobilisation (Ravi et al. 2004), but has not been considered in this study. Seasonal or short-term changes in precipitation may render some regions incapable to emit dust, while others may become more productive than in the annual mean. Also, different soil properties such as suspendibility have a large influence on the amount of dust emitted from a certain area, and lead to well-known "hot-spots" of dust emission (Goudie and Middleton 2001; Koren and Kaufman 2004). In addition, seasonality in vegetation cover is not accounted for, as the data set by DeFries and Townshend (1994) was generated from one year of averaged NDVI (normalised difference vegetation index) satellite data only. The largest part of our potential source area is however bare soil and should not show significant seasonal vegetation changes.

Physical properties of the dust itself, such as size spectra, have not been taken into account here. Size spectra are important for the dry deposition of heavier particles, and the ageing of mineral aerosol during transport. Typical settling velocities of long-range transported mineral aerosol are in the range of 0.001–0.02 $m\,s^{-1}$, but may be significantly higher for large particles (Duce et al. 1991). Hence, for longer transportation times, such as during phase M1 and M2, dry deposition of larger particles is likely to be more relevant than for the short transport paths observed during phase M3 and the October event. These influences are only taken into account in a very qualitative way in this study.

Wet deposition of mineral aerosol at Piz Zupó was identified with a relative humidity criterion, and confirmed by nearby precipitation with respect to timing. However, undisputable evidence that scavenging of mineral aerosol actually is taking place would require additional information. Aerosol measurements from the nearby JFJ observatory provided valuable but limited evidence, as the two locations may experience a different timing of the dust events. In addition, as noted above, high aerosol concentrations are a necessary but insufficient criterion for the formation of dust layers in snow, as precipitation is needed for efficient deposition (Osada et al. 2004). Finally, only detailed analyses of precipitation chemistry could provide insight into the wet removal process that is actually taking place during a dust event (Schwikowski et al. 1995). Given such detailed data were available, they would allow for an interesting validation study of our back-trajectory methodology.

The qualitative methodology presented here can be considered as a first-order Lagrangian dust transport model. It could be extended to a full dust transport model by including parameterisations of further dust-related processes, in particular convection, coagulation, and gravitational settling of dust particles. Currently, however, it is essential that all estimates of source regions and transport pathways of dust in this study be interpreted as "potential" source regions and "estimates" for transport pathways and processes along them. Future applications of such a refined Lagrangian transport model could include modelling the transport of other aerosols, for instance from urban plumes, or biogenic substances, such as pollen.

A.8 Conclusions

In an ice core recovered from the high-accumulation glacial site Piz Zupó in the Swiss Alps, two large dust events were identified that occurred in March and October 2000. The two dust events were characterised by pronounced changes in chemical composition compared to other sections of the ice core. A particularly noteworthy chemical difference existed also between the two dust events, namely exceptionally high concentrations of methanesulphonic acid (MSA) in the M3 case. The two identified periods were studied in-depth by means of a new method based on three-dimensional kinematic back-trajectories combined with objective selection criteria. The aims of this study were (i) to understand how the mobilisation area, transport pathway, and depositional conditions of the Saharan dust contributed to the chemical signatures observed in the ice core, and (ii) to investigate if different evolutions of the synoptic flow can lead to similar occurrences of dust layers in an Alpine ice core.

For October 2000, the extended back-trajectory analysis indicated that intense dust mobilisation occurred along a region stretching from northern Mauritania across Algeria into Tunesia during 11–15 October. Subsequently, mobilised dust was rapidly transported across the Mediterranean towards the Alps from southerly directions, and deposited at Piz Zupó during the following 2-3 days (Table A.1).

In March 2000, three phases of Saharan dust mobilisation and deposition at Piz Zupó were identified. During the first minor phase probably all dust rained out before arriving at the glacier. During the second phase, dust was transported during 9 days along a rather exceptional pathway across the eastern North Atlantic, approaching the Alps from northerly directions. Only a few days later a third dust transport phase occurred, bringing Saharan air masses directly from southerly directions across the Mediterranean towards the Alps, similar to the October event (Tab. A.1).

Our main conclusions from the analysis of these two dust events are as follows:

1. Major differences in the chemical signature of the two dust events preserved in the ice core were shown to be related to different process histories during transport rather than source regions. High concentrations of MSA in March 2000 were most likely caused by aerosol scavenging during the passage of the dust through the marine boundary layer over areas of high phytoplankton productivity in the western Mediterranean. In October 2000, despite a similar transport pathway, the potential for MSA scavenging was reduced due to seasonally low phytoplankton activity. In March, pollution sources that were encountered during the transport probably led to highly enhanced concentrations of NH_4^+ and NO_3^-.

2. The transport patterns of dust from the Saharan desert to the Swiss Alps can vary substantially from case to case. Even within a single large event identified in the ice core, they may change drastically from one day to another. It is thus not possible to identify one single meteorological situation which is typically associated with the transport of dust towards central Europe. This underlines the variability of dust transport into the Alpine region. Particularly strong dust events recorded at our study site and rapid transfer of Saharan dust to the Alps can be induced by the presence of large upper-level PV streamers that reach to northern Africa. In many cases, this flow configuration can also be associated with heavy precipitation at the Alpine southside, and hence favour the wet deposition of dust.

3. Potential source regions of dust were identified in the Algerian and Libyan deserts in the case of dust transport directly from the south, while mobilisation areas for transport across the Atlantic were centrered around Mauritania. It is currently not possible to clearly corroborate the source area identification from the chemical composition of the dust, as no definite chemical tracer is readily available. Visible satellite imagery however suggests that the potential mobilisation regions identified from the Lagrangian analysis are generally in good agreement with the actual source regions of dust.

4. The novel dust-transport analysis based on back-trajectories and objective criteria for dust mobilisation and deposition successfully captured the transport pathways of Saharan dust, and established a direct link between atmospheric circulation features and the chemical signal in the ice core. Dust events having lost most of their dust load due to wet deposition before reaching the arrival site, but which erroneously would have been identified by examining back-trajectories in a conventional way, could clearly be rejected with the methodology applied here. Additional observational evidence, such as satellite images, air chemistry measurements, and precipitation data, proved to be very valuable in order to back up the results of the advanced trajectory study.

From our study, it becomes evident that in order to interpret the complex chemical signal of dust deposited in an ice core it is essential to examine the full transport sequence of dust mobilisation, transport, and deposition. The new back-trajectory methodology, which incorporates additional meteorological information combined with further observational evidence proved to be very valuable for this kind of analysis.

Acknowledgements

Chlorophyll *a* data were provided by the SeaWiFS Project, NASA/Goddard Space Flight Center and Orbimage. ARPA Lombardia is acknowledged for access to the precipitation data. MeteoSwiss is acknowledged for access to the ECMWF data and the satellite imagery. This project was partly funded by the Swiss NCCR Climate programme.

Appendix B

Lagrangian moisture source diagnostic details

B.1 Trajectory calculation details

Table B.1 contains details of the back-trajectory calculation for the 30 selected months. The table contains from left to right the monthly NAO index (Hurrell 1995), the case name (format YYYYMM), days, maximum possible number of trajectories, calculation errors, precipitating trajectories at arrival (RH$\geq$80%), uptaking trajectories (at least one identified moisture source), and the ratio of precipitating to uptaking trajectories.

B.2 Diagnosed evaporation areas

Figure B.1 shows the diagnosed evaporation areas. This can be considered as the evaporation into uptaking air parcels without considering later rainout (no attribution according to Sec. 3.2). Note that in comparison with Fig. 3.7, evaporation areas are more widespread, and extend further into the subtropics and the western Atlantic. Note also that evaporation into bypassing trajectories is diagnosed in the Pacific. Subsequent rainout during transport then reduces the effective precipitation contribution of these more distant sources.

B.3 Uptakes above the boundary layer

Figure B.2 shows the locations of identified moisture uptake in air parcels that were located above the boundary layer at the time of uptake. The areas of above-boundary layer (ABL) uptakes agrees roughly with the impression gained from the below-boundary layer uptakes (comp. Fig. 3.7). It can however be noted that the ABL uptakes show a maximum at the slope of Greenland, and to a larger part take place over land ar-

Table B.1: Back-trajectory calculation setup and summary. See text for details.

NAO Index	Case	Days	Trajectories (#)	Errors (#)	Precipitating (%)	Uptaking (%)	Perc. of Upt. (%)
NAO positive (3.89±0.60)							
4.8	199702	28	668080	0	1.14	1.11	97.4
4.7	199002	28	668080	1	0.97	0.90	92.9
4.3	199001	31	739660	0	1.95	1.85	94.9
4.1	198902	28	668080	0	0.48	0.45	94.5
4.1	198401	31	739660	3	1.14	1.10	96.6
3.9	200002	29	691940	0	1.07	1.02	95.3
3.7	197401	31	739660	1	2.42	2.34	96.5
3.2	199301	31	739660	5	1.03	1.00	96.4
3.1	199502	28	668080	1	1.19	1.13	94.9
3.0	199312	31	739660	0	1.65	1.58	95.9
NAO neutral (0.01±0.30)							
0.5	198101	31	739660	3	1.29	1.20	93.6
0.4	198312	31	739660	4	1.45	1.39	95.4
0.2	199112	31	739660	23	2.85	2.77	97.2
0.2	196801	31	739660	3	1.35	1.27	94.5
0.1	198802	29	691940	6	2.38	2.25	94.4
-0.1	199302	28	668080	18	2.75	2.69	97.8
-0.2	196412	29	691940	21	1.96	1.89	96.4
-0.3	199712	31	739660	3	3.90	3.69	94.6
-0.3	196501	31	739660	2	2.15	2.04	94.8
-0.4	199402	28	668080	0	2.35	2.28	97.3
NAO negative (-5.00±0.83)							
-4.0	196902	28	668080	3	3.32	3.17	95.3
-4.3	196901	31	739660	4	2.28	2.11	92.5
-4.4	195901	31	739660	7	1.21	1.15	94.8
-4.5	196601	31	739660	2	1.66	1.58	95.4
-4.6	198701	31	739660	63	6.69	6.42	96.0
-5.1	196502	28	668080	28	5.89	5.64	95.8
-5.1	199612	31	739660	44	4.53	4.31	95.1
-5.2	199512	31	739660	3	3.30	3.17	96.3
-5.8	196112	31	739660	2	2.46	2.32	94.1
-7.0	196301	31	739660	23	2.91	2.78	95.6

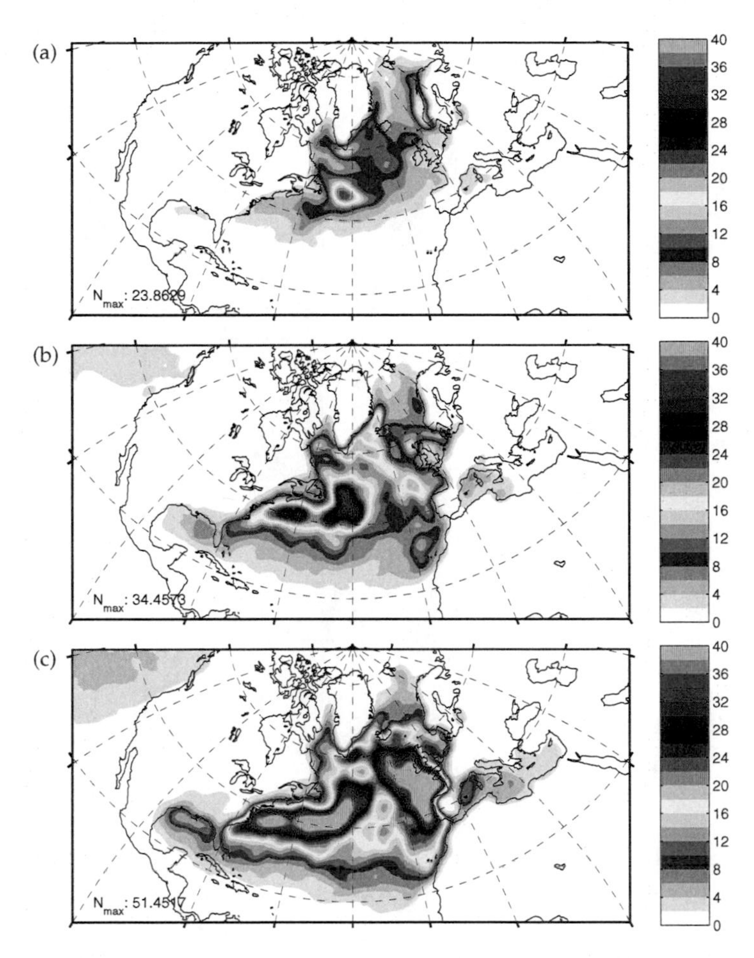

Figure B.1: Phase mean moisture sources (no attribution) for (a) NAO+, (b) NAO=, (c) NAO− months. Uptake locations show the evaporation into air parcels on the way to Greenland. Units are mm evaporation integrated over 10^4 km^2.

eas. While the former could possibly be related to orographically forced convection, the latter could be an indication that the boundary-layer parametrisation of the ECMWF model has problems in predicting an accurate winter-time boundary layer height over land. The fact that both, below and above boundary-layer uptakes, match to a large degree confirms that the identified moisture sources are a robust finding, and only may depend quantitatively on the ECMWF boundary layer parametrisation. This finding was also supported by the inspection of vertical cross-sections for several selected days in December 1997 (not shown). Nevertheless, further investigation into the reasons for and circumstances of uptakes above the boundary layer could give valuable insight into the performance of the moisture source diagnostic, in particular during summer months.

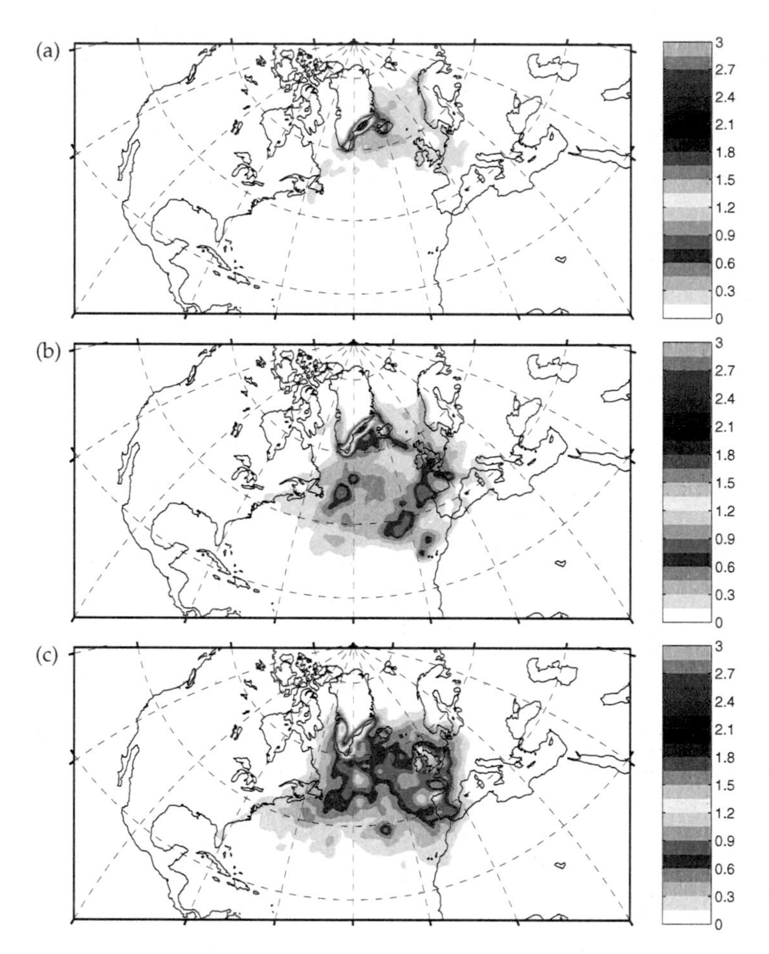

Figure B.2: Phase mean moisture uptake above the boundary layer for (a) NAO+, (b) NAO=, (c) NAO− months. Uptake locations show the contribution to precipitation in Greenland. Units are mm precipitation contribution integrated over 10^4 km^2.

B.4 Regionalised moisture source regions

In order to attain a more regional view of the influence of different moisture source regions, the ice sheet was subdivided into 5 different arrival regions (Fig. B.3). The subdivision into a northern, central east, central, central west, and southern sector broadly follows the regions defined by Bromwich et al. (1999). The moisture sources for the different regions show different sensitivities to the NAO (Fig. B.4). Note that since each panel is scaled to the maximum value of the respective region and NAO phase, no direct quantitative comparison is possible from this figure. However, structural changes are clearly highlighted, and allow for distinguishing regions of very pronounced changes (e.g. Fig. B.4a, c) from such where changes with NAO phase are weaker (e.g. Fig. B.4d, e). This sectorised view hence complements the Lagrangian forward projection perspective of, for example, source region latitude changes (comp. Fig. 3.9).

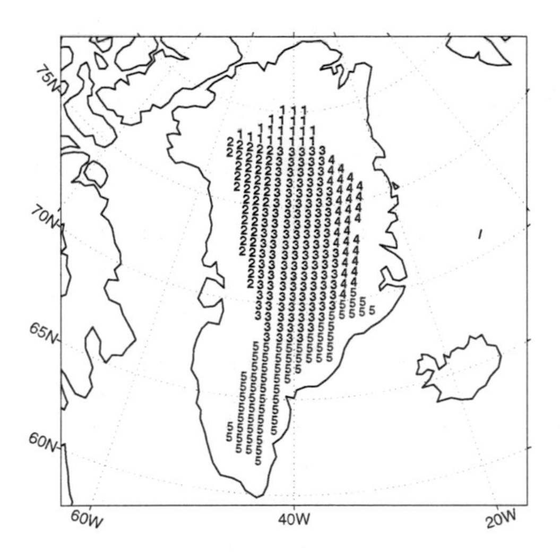

Figure B.3: Subdivision of the Greenland ice sheet into five regions, following the sectors defined by Bromwich et al. (1999). 1: Northern, 2: Central west, 3: Central, 4: Central east, 5: Southern sector.

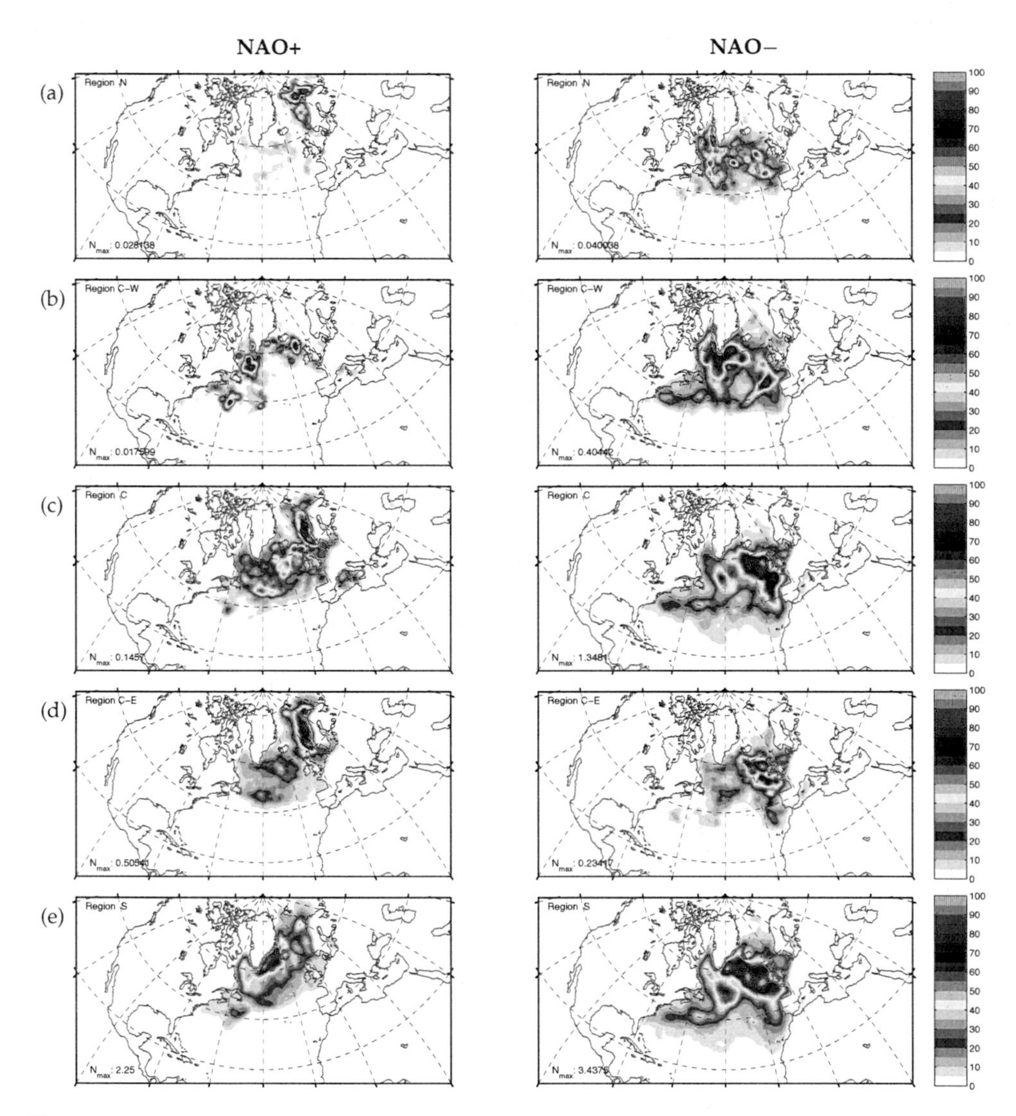

Figure B.4: Phase mean moisture sources for NAO+ and NAO− months, subdivided into 5 arrival regions over the Greenland ice sheet. (a) Northern, (b) Central west, (c) Central, (d) Central east, (e) Southern sector. Uptake locations show the contribution to precipitation in Greenland. Units are percent of the maximum value in each panel.

Appendix C

MCIM model setup and adjustments

C.1 The Mixed-Cloud Isotope Model MCIM

The MCIM model is a one-dimensional Rayleigh fractionation model that calculates the isotopic fractionation of water vapour along an idealised pressure-temperature ($p - T$) trajectory from an oceanic source to the arrival at a high-latitude precipitation site. The initialisation of the model from surface conditions at the evaporation site is based on a global closure equation, which describes the isotopic composition of water vapour depending on relative humidity (RH), sea surface temperature (SST), and wind velocity at an evaporation site (Merlivat and Jouzel 1979). Later, Jouzel and Merlivat (1984) included kinetic fractionation effects during snow formation in the MCIM model which occur during supersaturated conditions in the cloud with respect to ice. This supersaturation function $S_i = 1.02 - 0.0038 \cdot T$ describes the fractionation during the Bergeron-Findeisen process, and acts towards reducing the d-excess with decreasing temperatures. However, this function is hardly constrained by observations. Ciais and Jouzel (1994) extended the model with a mixed-phase microphysical parametrisation that allows the coexistence of ice and liquid droplets in the cloud during a certain temperature range. Further model-internal fitting parameters concern the rate of liquid that is retained in the cloud versus the precipitation ($C_{1,prec}$ and $C_{2,prec}$, Ciais and Jouzel 1994).

The model was tuned to specific 'climatological' source and arrival conditions, which can be strongly different on a synoptic time scale. The (subtropical) moisture source temperature is assumed to have about 15–20°C. Temperature at the arrival location of the cloud is calculated from the empirical relation $T_{arr} = T_{sfc} \cdot 0.67 - 1.2$ (°C) (Jouzel and Merlivat 1984), with an annual mean surface temperature $T_{sfc} = -30$°C. The tuning of the model will hence comprise the effects of the relation between T_{sfc} and T_{arr} via the microphysical processes related to precipitation.

The model's global closure approach for calculating the initial isotopic composition of the water vapour is probably not valid on a local scale, and biased towards too high SSTs, as has been shown by Jouzel and Koster (1996). These authors have therefore recommended to use the isotope composition of water vapour at the lowest level of an isotope GCM for initialisation.

Technically, the fractionation calculation begins with evaporation according to the SST, RH, and wind velocity conditions at the moisture source. Then, the air parcel is lifted with an uniform lapse rate until condensation is reached (RH=100%). Then, driven by further adiabatic and radiative cooling, and proceeding linearly up to the arrival location, Rayleigh fractionation is calculated. Here, we apply the MCIM model to diagnosed condensation and arrival temperatures and pressures. As these cannot be given to the model, a different approach, based on modified lapse rates (V. Masson-Delmotte, *pers. comm.*, 2005) has been chosen.

C.2 Adjusted lapse-rate calculation

The lapse-rate calculation was changed so that the condensation onset occurred at the condensation temperatures diagnosed here. To this end, the relative humidity and the lapse rate were modified, so that when an air parcel is lifted from the surface in MCIM, it cools according to the adjusted lapse rate, and cloud condensation level (RH=100%) is reached exactly at the specified T, p conditions.

Input values of the calculation are the 2 m air temperature T_0, the temperature at condensation level T_c, and the pressure at condensation level p_c. Output values are the relative humidity at the surface RH_0, and the adjusted lapse rate lr. In detail, the calculation of an adjusted lapse rate proceeds along the following steps:

1. Calculate the saturation vapour pressure e_c^* for the condensation temperature T_c:

$$e_c^* = 6.1078 \cdot \exp\left(a \frac{T - 273.16}{T - b}\right) \text{ (hPa)} \tag{C.1}$$

   where for ice a=21.8745584 and b=7.66, for water a=17.2693882 and b=35.86.

2. Find a condensation altitude z_c which corresponds to the diagnosed condensation pressure p_c assuming a US standard atmosphere.

3. Calculate the adjusted lapse rate lr from the vertical temperature gradient between the surface and condensation level:

$$lr = \frac{T_c - T_0}{z_c} \text{ K km}^{-1} \tag{C.2}$$

4. Calculate the relative humidify at the surface RH_0 corresponding to the water vapour pressure at the surface e_0^* using Eq. (C.1):

$$\text{RH}_0 = e_c^*/e_0^*. \tag{C.3}$$

Bibliography

Adams, J. W., D. Rodriguez, and R. A. Cox. 2005. The uptake of SO2 on Saharan dust: a flow tube study. *Atmos. Chem. Phys.* **5**, 2679–2689.

Aggarwal, P. K., M. A. Dillon, and A. Tanweer. 2004. Isotope fractionation at the soil-atmosphere interface and the 18O budget of atmospheric oxygen. *Geophys. Res. Lett.* **31**, L14202. 10.1029/2004GL019945.

Alpert, P., and E. Ganor. 1993. A jet-stream associated heavy dust storm in the western Mediterranean. *J. Geophys. Res.* **98**, 7339–7349.

Andersson, E., P. Bauer, A. Beljaars, F. Chevallier, E. Hólm, M. Janisková, P. Kallberg, G. Kelly, P. Lopez, A. McNally, E. Moreau, A. J. Simmons, J.-N. Thépaut, and A. M. Tompkins. 2005. Assimilation and modeling of the atmospheric hydrological cycle in the ECMWF forecasting system. *BAMS* **82**, 387–402.

Ansmann, A., J. Bösenberg, A. Chaikovsky, A. Comerón, S. Eckhardt, R. Eixmann, V. Freudenthaler, P. Ginoux, L. Komguem, H. Linné, M. A. L. Márquez, V. Matthias, I. Mattis, V. Mitev, D. Müller, S. Music, S. Nickovic, J. Pelon, L. Sauvage, P. Sobolewsky, M. K. Srivastava, A. Stohl, O. Torres, G. Vaughan, U. Wandinger, and M. Wiegner. 2003. Long-range transport of Saharan dust to northern Europe: The 11–16 October 2001 outbreak observed with EARLINET. *J. Geophys. Res.* **108**, 4783. 10.1029/2003JD003757.

Appenzeller, C., and H. C. Davies. 1992. Structure of stratospheric intrusions into the troposphere. *Nature* **358**, 570–572.

Appenzeller, C., H. C. Davies, and W. A. Norton. 1996. Fragmentation of stratospheric intrusions. *J. Geophys. Res.* **101**, 1435–1456.

Appenzeller, C., J. Schwander, S. Sommer, and T. Stocker. 1998a. The North Atlantic Oscillation and its imprint on precipitation and ice accumulation in Greenland. *Geophys. Res. Lett.* **25**, 1939–1942.

Appenzeller, C., T. Stocker, and M. Anklin. 1998b. North Atlantic Oscillation dynamics recorded in Greenland ice cores. *Science* **282**, 446–449.

Araguás-Araguás, L., K. Fröhlich, and K. Rozanski. 2000. Deuterium and oxygen-18 isotope composition of precipitation and atmospheric moisture. *Hydrol. Process.* **14**, 1341–1355.

Avila, A., I. Queralt-Mitjans, and M. Alarcon. 1997. Mineralogical composition of African dust delivered by red rains over northeastern Spain. *J. Geophys. Res.* **102**, 21977–21996.

Aymoz, G., J.-L. Jaffrezo, V. Jacob, A. Colomb, and C. George. 2004. Evolution of organic and inorganic components of aerosol during a Saharan dust episode observed in the French Alps. *Atmos. Chem. Phys.* **4**, 2499–2512.

Barkan, J., P. Alpert, H. Kutiel, and P. Kishcha. 2005. Synoptics of dust transportation days from Africa toward Italy and central Europe. *J. Geophys. Res.* **110**, D07208.

Barlow, L. K., J. C. Rogers, M. C. Serreze, and R. G. Barry. 1997. Aspects of climate variability in the North Atlantic sector: Discussion and relation to the Greenland Ice Sheet Project 2 high-resolution isotopic signal. *J. Geophys. Res.* **102**, 26333–26344.

Barlow, L. K., J. W. C. White, R. G. Barry, J. C. Rogers, and P. M. Grootes. 1993. The North Atlantic Oscillation signature in Deuterium and Deuterium excess signals in the Greenland ice sheet project 2 ice core, 1840-1970. *Geophys. Res. Lett.* **24**, 2901–2904.

Bergametti, G., L. Gomes, C. G., P. Rognon, and M.-N. le Coustumer. 1989. African dust observed over Canary Islands - Source-regions identification and transport pattern for some summer situations. *J. Geophys. Res.* **94**, 14855–14864.

Blattmann-Singh, M. 2005. *Durchführung idealisierter Simulationen im regionalen Klimamodell CHRM mit einem Wasserdampftracer*. B.sc. thesis. IAC, ETH Zürich.

Bleck, R., and C. Mattocks. 1984. A preliminary analysis of the role of potential vorticity in Alpine lee cyclogenesis. *Contrib. Atmos. Phys.* **57**, 357–368.

Bonasoni, P., P. Cristofanelli, F. Calzolari, U. Bonafè, F. Evangelisti, A. Stohl, S. Zauli, R. van Dingenen, T. Colombo, and Y. Balkanski. 2004. Aerosol-ozone correlations during dust transport episodes. *Atmos. Chem. Phys.* **4**, 1201–1215.

Bosc, E., A. Bricaud, and D. Antoine. 2004. Seasonal and interannual variability in algal biomass and primary production in the Mediterranean Sea, as derived from 4 years of SeaWiFS observations. *Global Biogeochem. Cycles*. 10.1029/2003GB002034.

Bosilovich, M. 2002. On the vertical distribution of local and remote sources of water for precipitation. *Meteorol. Atmos. Phys.* **80**, 31–41.

Bosilovich, M., and S. Schubert. 2002. Water vapour tracers as diagnostics of the regional hydrologic cycle. *J. Hydrometeorol.* Pp. 149–165.

Bosilovich, M. G., Y. C. Sud, S. D. Schubert, and G. K. Walker. 2003. Numerical simulation of the large-scale North American monsoon water sources. *J. Geophys. Res.* **108**, 8614.

Bricaud, A., E. Bosc, and D. Antoine. 2002. Algal biomass and sea surface temperature in the Mediterranean Basin Intercomparison of data from various satellite sensors, and implications for primary production estimates. *Remote Sens. Environ.* **81**, 163–178.

Bromwich, D. H., Q. Chen, Y. Li, and R. I. Cullather. 1999. Precipitation over Greenland and its relation to the North Atlantic Oscillation. *J. Geophys. Res.* **104**, 22103–22115.

Brown, J., and I. Simmonds. 2004. Sensitivity of the δ^{18}O-temperature relationship to the distribution of continents. *Geophys. Res. Lett.* **31**, L09208.

Brubaker, K. L., P. A. Dirmeyer, A. Sudradjat, B. S. Levy, and F. Bernal. 2001. A 36-yr climatological description of the evaporative sources of warm-season precipitation in the Mississippi river basin. *J. Hydromet.* **2**, 537–557.

Budyko, M. I. 1974. *Climate and Life*. Academic Press, San Diego.

Cappa, C. D., M. B. Hendricks, D. J. DePaolo, and R. C. Cohen. 2003. Isotopic fractionation of water during evaporation. *J. Geophys. Res.* **108**, 4525.

Charles, R. D., D. Rind, J. Jouzel, R. D. Koster, and R. G. Fairbanks. 1994. Glacial-interglacial changes in moisture sources for Greenland: Influences on the ice core record of climate. *Science* **263**, 508–511.

Christensen, J. H., and O. B. Christensen. 2003. Severe summertime flooding in Europe. *Nature* **421**, 805–806.

Ciais, P., and J. Jouzel. 1994. Deuterium and oxygen 18 in precipitation: Isotopic model, including mixed cloud processes. *J. Geophys. Res.* **99**, 16793–16803.

Ciais, P., J. W. C. White, J. Jouzel, and J. R. Petit. 1995. The origin of present-day Antarctic precipitation from surface snow deuterium excess data. *J. Geophys. Res.* **100**, 18917–18927.

Claquin, T., M. Schulz, and Y. J. Balkanski. 1999. Modeling the mineralogy of atmospheric dust sources. *J. Geophys. Res.* **104**, 22243–22256.

Collaud Coen, M., E. Weingartner, D. Schaub, C. Hueglin, C. Corrigan, S. Henning, M. Schwikowski, and U. Baltensperger. 2004. Saharan dust events at the Jungfraujoch: detection by wavelength dependence of the single scattering albedo and first climatology analysis. *Atmos. Chem. Phys.* **4**, 2465–2480.

Craig, H., and L. I. Gordon. 1965. Deuterium and oxygen 18 variations in the ocean and the marine atmosphere. Pp. 9–130. In *Stable isotopes in oceanographic studies and paleotemperatures*. E. Tongiorgi (ed.). Consiglio nazionale delle ricerche, Laboratorio di geologia nucleare, Pisa.

Dahl-Jensen, D., K. Mosegaard, N. Gundestrup, G. D. Clow, S. J. Johnsen, A. Hansen, and N. Balling. 1998. Past temperatures directly from the Greenland ice sheet. *Science* **282**, 268–271.

Dansgaard, W. 1964. Stable isotopes in precipitation. *Tellus* **16**, 436–468.

Dansgaard, W. e. a. 1993. Evidence for past climate from a 250-kyr ice-core record. *Nature* **364**, 218–220.

Davies, H. C. 1976. A lateral boundary formulation for multi-level prediction models. *Q. J. R. Mereorol. Soc.* **102**, 405–418.

DeFries, R. S., and J. R. G. Townshend. 1994. NDVI-derived land cover classification at global scales. *Int. J. Remote Sensing* **15**, 3567–3586.

Delaygue, G., V. Masson, J. Jouzel, R. D. Koster, and R. J. Healy. 2000. The origin of Antarctic precipitation: a modelling approach. *Tellus* **52B**, 19–36.

Delmotte, M., V. Masson, J. Jouzel, and V. I. Morgan. 2000. A seasonal deuterium excess signal at Law Dome, coastal eastern Antarctica: A southern ocean signature. *J. Geophys.Res.* **105**, 7187–7197.

Dickinson, R. E. 1984. *Modeling Evapotranspiration for 3-dimensional Global Climate Models*. Pp. 58–72. number 29 in *Geophysical Monographs*. American Geophysical Union.

Dirmeyer, P. A., and K. L. Brubaker. 1999. Contrasting evaporative moisture sources during the drought of 1988 and the flood of 1993. *J. Geophys. Res.* **104**, 19383–19397.

Druyan, L. M., and R. D. Koster. 1989. Sources of Sahel precipitation for simulated drought and rainy seasons. *J. Climate* **2**, 1438–1446.

Duce, R., P. S. Liss, J. T. Merrill, E. L. Atlas, P. Buatt-Ménard, B. B. Hicks, J. M. Miller, J. M. Prospero, R. Arimoto, T. M. Church, W. Ellis, J. N. Galloway, L. Hansen, T. D. Jickells, A. H. Knap, K. H. Reinhardt, B. Schneider, A. Soudine, J. J. Tokos, S. Tsunogai, R. Wollast, and M. Zhou. 1991. The atmospheric input of trace gas species to the world ocean. *Global Biogeochem. Cycles* **5**, 193–259.

DWD 1995. *Dokumentation des EM/DM Systems*. Deutscher Wetterdienst, Abteilung Forschung, Offenbach/Main.

Eckhardt, S., A. Stohl, H. Wernli, P. James, C. Forster, and N. Spichtinger. 2004. A 15-year climatology of warm conveyor belts. *J. Climate* **17**, 218–237.

Eichler, A., M. Schwikowski, H. W. Gäggeler, V. Furrer, H.-A. Synal, J. Beer, M. Saurer, and M. Funk. 2000. Glaciochemical dating of an ice core from upper Grenzgletscher (4200 m a.s.l). *J. Glaciol.* **46**, 507–515.

EPICA community members 2004. Eight glacial cycles from an Antarctic ice core. *Nature* **429**, 623–628.

Federer, B., B. Thalmann, and J. Jouzel. 1982a. Stable isotopes in hailstones. Part II: Embryo and hailstone growth in different storms. *J. Atmos. Sci.* **39**, 1336–1355.

Federer, B., N. Brichet, and J. Jouzel. 1982b. Stable isotopes in hailstones. Part I: The isotopic cloud model. *J. Atmos. Sci.* **39**, 1323–1335.

Fischer, M. J., and K. Sturm. 2006. REMOiso forcing for the iPILPS Phase 1 experiments and the. *Hydrol. Process.* **51**, 73–89. 10.1016/j.gloplacha.2005.12.006.

Franzén, L. G., M. Hjelmroos, P. Kållberg, A. Rapp, J. O. Mattsson, and E. Brorström-Lundén. 1995. The Saharan dust episode of south and central Europe, and northern Scandinavia, March 1991. *Weather* **50**, 313–318.

Frei, C., J. H. Christensen, M. Déqué, D. Jacob, R. G. Jones, and P. L. Vidale. 2003. Daily precipitation statistics in regional climate models: Evaluation and intercomparison for the European Alps. *J. Geophys. Res.* **108**, 4124.

Fukutome, S., C. Prim, and C. Schär. 2001. The role of soil states in medium-range weather predictability. *Nonlin. Process. Geophys.* **8**, 373–386.

Gat, J. R. 1996. Oxygen and hydrogen isotopes in the hydrologic cycle. *Annu. Rev. Earth Planet. Sci.* **24**, 225–262.

Gat, J. R., B. Klein, Y. Kushnir, W. Roether, H. Wernli, R. Yam, and A. Shemesh. 2003. Isotope composition of air moisture over the Mediterranean Sea: an index of air-sea interaction pattern. *Tellus* **55B**, 953–965.

Gedzelman, S. D., and J. R. Lawrence. 1982. The isotopic composition of cyclonic precipitation. *J. Appl. Meteorol.* **21**, 1385–1404.

Gedzelman, S. D., and J. R. Lawrence. 1990. The isotopic composition of precipitation from two extratropical cyclones. *Mon. Wea. Rev.* **118**, 495–509.

Gedzelman, S. D., and R. Arnold. 1994. Modeling the isotopic composition of precipitation. *J. Geophys. Res.* **99**, 10455–10472.

Gedzelman, S. D., J. M. Rosenbaum, and J. R. Lawrence. 1989. The megalopolitan snowstorm of 11-12 February 1983: Isotopic composition of the snow. *J. Atmos. Sciences* **46**, 1637–1649.

Gedzelman, S., J. Lawrence, J. Gamache, M. Black, E. Hindman, R. Black, J. Dunion, H. Willoughby, and X. P. Zhang. 2003. Probing hurricanes with stable isotopes of rain and water vapor. *Mon. Wea. Rev.* **131**, 1112–1127.

Goudie, A. S., and N. J. Middleton. 2001. Saharan dust storms: nature and consequences. *Earth-Sci. Rev.* **56**, 179–204.

Hamonou, E., P. Chazette, D. Balis, F. Dulac, X. Schneider, E. Galani, G. Ancellet, and A. Papayannis. 1999. Characterization of the vertical structure of Saharan dust export to the Mediterranean basin. *J. Geophys. Res.* **104**, 22257–22270.

Hanna, E., P. Valdes, and J. McConnell. 2001. Patterns and variations of snow accumulation over Greenland, 1979-98, from ECMWF analyses, and their verification. *J. Climate* **14**, 3521–3535.

Hecht, M. W., W. R. Holland, and P. J. Rasch. 1995. Upwind-weighted advection schemes for ocean tracer transport: An evaluation in a passive tracer context. *J. Geophys. Res.* **100**, 20763–20778.

Heck, P., D. Luethi, H. Wernli, and C. Schaer. 2001. Climate impacts of European-scale anthropogenic vegetation changes: A sensitivity study using a regional climate model. *J. Geophys. Res.* **106**, 7817–7835.

Helsen, M. M. 2005. *On the interpretation of stable isotopes in Antarctic precipitation.* PhD Thesis, Utrecht University.

Helsen, M. M., R. S. W. van de Wal, M. R. van den Broeke, E. R. T. Kerstel, V. Masson-Delmotte, H. A. J. Meijer, C. H. Reijmer, and M. P. Scheele. 2004. Modelling the isotopic composition of snow using backward trajectories: a particular precipitation event in Dronning Maud Land, Antarctica. *Ann. Glaciol.* **39**, 293–299.

Helsen, M. M., R. S. W. van de Wal, M. R. van den Broeke, V. Masson-Delmotte, H. A. J. Meijer, M. P. Scheele, and M. Werner. 2005a. Modelling the isotopic composition of Antarctic snow using backward trajectories: Simulation of snow pit records. *J. Geophys. Res.* Pp. submitted.

Helsen, M. M., R. W. van de Wal, M. van den Broeke, D. van As, H. Meijer, and C. Reijmer. 2005b. Oxygen isotope variability in snow from western Dronning Maud Land, Antarctica and its relation to temperature. *Tellus* **57B**, 423–435.

Henning, S., E. Weingartner, M. Schwikowski, H. W. Gaeggeler, R. Gehrig, K.-P. Hinz, A. Trimborn, B. Spengler, and U. Baltensperger. 2003. Seasonal variation of water-soluble ions of the aerosol at the high-alpine site Jungfraujoch (3580 m asl). *J. Geophys. Res.* **108**, 4030. 10.1029/2002JD002439.

Hinz, K.-P., A. Trimborn, E. Weingartner, S. Henning, U. Baltensperger, and B. Spengler. 2005. Aerosol single particle composition at the Jungfraujoch. *J. Aerosol Sci.* **36**, 123–145.

Hoerling, M., and A. Kumar. 2003. The perfect ocean for drought. *Science* **299**, 691–694.

Hoffmann, G., J. Jouzel, and V. Masson. 2000. Stable water isotopes in atmospheric general circulation models. *Hydrol. Process.* **14**, 1385–1406.

Hoffmann, G., M. Werner, and M. Heimann. 1998. Water isotope module of the ECHAM atmospheric general circulation model: A study on timescales from days to several years. *J. Geophys. Res.* **103**, 16871–16896.

Holzer, M., T. M. Hall, and R. B. Stull. 2005. Seasonality and weather-driven variability of transpacific transport. *J. Geophys. Res.* **110**, D23103.

Houghton, J. T., Y. Ding, D. J. Griggs, M. Noguer, P. J. van der Linden, and D. Xiaosu. (Eds.) 2001. *Climate Change 2001: The Scientific Basis. Contribution of Working Group I to the Third Assessment Report of the Intergovernmental Panel on Climate Change (IPCC).* Cambridge University Press, UK.

Houze Jr., R. A. 1993. *Cloud Dynamics.* Academic Press, San Diego.

Huebert, B. J., B. W. Blomquist, J. E. Hare, C. W. Fairall, J. E. Johnson, and T. S. Bates. 2004. Measurement of the air-sea DMS flux and transfer velocity using eddy correlation. *Geophys. Res. Lett.* **31**, L23113. 10.1029/2004GL021567.

Hurrell, J. W. 1995. Decadal trends in the North Atlantic oscillation: Regional temperatures and precipitation. *Science* **269**, 676–679.

Hurrell, J. W., Y. Kushir, G. Ottersen, and M. Visbeck. 2003. *The North Atlantic Oscillation.* number 134 in *Geophysical Monographs.* American Geophysical Union, Washington, USA.

Jacobsen, I., and E. Heise. 1982. A new economic method for the computation of the surface temperature in numerical models. *Beitr. Phys. Atmos.* **55**, 128–141.

James, P., A. Stohl, N. Spichtinger, S. Eckhardt, and C. Forster. 2004. Climatological aspects of the extreme European rainfall of August 2002 and a trajectory method for estimating the associated evaporative source regions. *NHESS* **4**, 733–746.

Johnsen, S. J., W. Dansgaard, and J. W. C. White. 1989. The origin of Arctic precipitation under present and glacial conditions. *Tellus* **41B**, 452–468.

Jones, C., N. Mahowald, and C. Luo. 2003. The role of easterly waves on African desert dust transport. *J. Climate* **16**, 3617–3628.

Jordan, C. E., J. E. Dibb, B. E. Anderson, and H. E. Fuelberg. 2003. Uptake of nitrate and sulfate on dust aerosols during TRACE-P. *J. Geophys. Res.* **108**, 8817. 10.1029/2002JD003101.

Joussaume, J., R. Sadourny, and J. Jouzel. 1984. A general circulation model of water isotope cycles in the atmosphere. *Nature* **311**, 24–29.

Joussaume, S., R. Sadourny, and C. Vignal. 1986. Origin of precipitating water in a numerical simulation of July climate. *Ocean-Air Interact.* **1**, 43–56.

Jouzel, J., and L. Merlivat. 1984. Deuterium and oxygen 18 in precipitation, modelling of the isotopic effects during snow formation. *J. Geophys. Res.* **89**, 11749–11757.

Jouzel, J., and R. D. Koster. 1996. A reconsideration of the inital conditions used for stable water isotope models. *J. Geophys. Res.* **101**, 22933–22938.

Jouzel, J., L. Merlivat, and E. Roth. 1975. Isotopic study of hail. *J. Geophys. Res.* **80**, 5015–5030.

Jouzel, J., R. B. Alley, K. M. Cuffey, W. Dansgaard, P. Grootes, G. Hoffmann, S. J. Johnsen, R. D. Koster, D. Peel, C. A. Shuman, M. Stievenard, M. Stuiver, and J. White. 1997. Validity of the temperature reconstruction from water isotopes in ice cores. *J. Geophys. Res.* **102**, 26471–26487.

Kavanaugh, J. L., and K. M. Cuffey. 2003. Space and time variation of d18O and dD in Antarctic precipitation revisited. *Global Biogeochem. Cycles* **17**, 1017. 10.1029/2002GB001910.

Keil, C., H. Volkert, and D. Majewski. 1999. The Oder flood in July 1997: Transport routes of precipitable water diagnosed with an operational forecast model. *Geophys. Res. Lett.* **26**, 235–238.

Kessler, E. 1969. *On the distribution and continuity of water substance in atmospheric circulations.* number 10 in *Meteorological Monographs*. American Geophysical Union.

Knippertz, P., A. H. Fink, A. Reiner, and P. Speth. 2003. Three late summer/early autumn cases of tropical-extratropical interactions causing precipitation in northwest Africa. *Mon. Wea. Rev.* **131**, 116–135.

Koren, I., and Y. J. Kaufman. 2004. Direct wind measurements of Saharan dust events from Terra and Aqua satellites. *Geophys. Res. Lett.* **31**, L06122. 10.1029/2003GL019338.

Koster, R. D., J. Jouzel, R. J. Suozzo, and G. L. Russell. 1992. Origin of July Antarctic precipitation and its influence on deuterium content: a GCM analysis. *Climate Dynamics* **7**, 195–203.

Koster, R., J. Jouzel, R. Souzzo, G. Russell, W. Broecker, D. Rind, and P. Eagleson. 1986. Global sources of local precipitation as determined by the NASA/GISS GCM. *Geophys. Res. Lett.* **13**, 121–124.

Krinner, G., and M. Werner. 2003. Impact of precipitation seasonality changes on isotopic signals in polar ice cores: a multi-model analysis. *Earth Plan. Sci. Lett.* **216**, 525–538.

Krinner, G., C. Genthon, and J. Jouzel. 1997. GCM analysis of local influences on ice core d signals. *Geophys. Res. Lett.* **24**, 2825–2828.

Lawrence, J. R., and S. D. Gedzelman. 1996. Low stable isotope ratios of tropical cyclone rains. *Geophys. Res. Lett.* **23**, 527–530.

Lawrence, J. R., S. D. Gedzelman, X. Zhang, and R. Arnold. 1998. Stable isotope ratios of rain and vapor in 1995 hurricanes. *J. Geophys. Res.* **103**, 11381–11400.

Leon, J.-F., D. Tanre, J. Pelon, Y. J. Kaufman, J. M. Haywood, and B. Chatenet. 2003. Profiling of a Saharan dust outbreak based on a synergy between active and passive remote sensing. *J. Geophys. Res.* **108**, 8575. 10.1029/2002JD002774.

Lin, S.-J., and R. B. Rood. 1996. Multidimensional flux form semi-Lagrangian transport schemes. *Mon. Wea. Rev.* **124**, 2046–2070.

Lin, Y.-L., R. D. Farley, and H. D. Orville. 1983. Bulk parameterization of the snow field in a cloud model. *J. Climate Appl. Meteor.* **22**, 1065–1092.

Loewe, E. 1936. The Greenland ice cap as seen by a meteorologist. *Quart. J. Roy. Meteorol. Soc.* **62**, 359–377.

Louis, J.-F. 1979. A parametric model of vertical eddy fluxes in the atmosphere. *Bound.-Layer Meteorol.* **17**, 187–202.

Louis, J.-F., M. Tiedtke, and J. F. Geleyn. 1982. A short history of the operational PBL parameterization at ECMWF. *Proceedings of the ECMWF Workshop on Boundary Layer Parameterizations.* ECMWF. Reading, UK. Pp. 59–79.

Lüthi, D., A. Cress, H. C. Davies, C. Frei, and C. Schär. 1996. Interannual variability and regional climate simulations. *Theor. Appl. Climatol.* **53**, 185–209.

Mahowald, N. M., and J.-L. Dufresne. 2004. Sensitivity of TOMS aerosol index to boundary layer height: Implications for detection of mineral aerosol sources. *Geophys. Res. Lett.* **31**, L03103. 10.1029/2003GL018865.

Majewski, D. 1991. The Europa-Model of the Deutscher Wetterdienst. *Proceedings of the ECMWF Workshop on Boundary Layer Parameterizations.* Vol. 2 of *ECMWF Proceedings of numerical methods in atmospheric models.* ECMWF. Reading, UK. Pp. 147–191.

Majoube, M. 1971. Fractionnement en oxygéne 18 et en deutérium entre l'eau et sa vapeur. *J. Chem. Phys.* **10**, 1423–1436.

Marsh, T. J., and R. B. Bradford. 2003. The floods of August 2002 in central Europe. *Weather* **58**, 168.

Martius, O., E. Zenklusen, C. Schwierz, , and H. C. Davies. 2006. Episodes of Alpine heavy precipitation with an overlying elongated stratospheric intrusion: A Climatology. *Int. J. Climatol.* **26**, 1149–1164. 10.1002/joc.1295.

Massacand, A. C., H. Wernli, and H. C. Davies. 1998. Heavy precipitation on the Alpine southside: An upper-level precursor. *Geophys. Res. Lett.* **25**, 1435–1438.

Masson-Delmotte, V., A. Landais, M. Stievenard, O. Cattani, S. Falourd, J. Jouzel, S. J. Johnsen, D. Dahl-Jensen, A. Sveinsbjornsdottir, J. W. C. White, T. Popp, and H. Fischer. 2005a. Holocene climatic changes in Greenland: Different deuterium excess signals at Greenland Ice Core Project (GRIP) and NorthGRIP. *J. Geophys. Res.* **110**, D14102.

Masson-Delmotte, V., J. Jouzel, A. Landais, M. Stievenard, S. J. Johnsen, J. W. C. White, M. Werner, A. Sveinbjornsdottir, and K. Fuhrer. 2005b. GRIP deuterium excess reveals rapid and orbital-scale changes in Greenland moisture origin. *Science* **309**, 118–121.

Mattis, I., A. Ansmann, D. Mueller, U. Wandinger, and D. Althausen. 2002. Dual-wavelength Raman lidar observations of the extinction-to-backscatter ratio of Saharan dust. *Geophys. Res. Lett.* **20**, 20. 10.1029/2002GL014721.

Mellor, G. L., and T. Yamada. 1974. A hierarchy of turbulent closure models for planetary boundary layers. *J. Atmos. Sci.* **31**, 1791–1806.

Merlivat, L., and J. Jouzel. 1979. Global climatic interpretation of the Deuterium-Oxygen 18 relationship for precipitation. *J. Geophys. Res.* **84**, 5029–5033.

Mladek, R., J. Barckicke, P. Binder, P. Bougeault, N. Brzovic, C. Frei, J. F. Geleyn, J. Hoffman, W. Ott, T. Paccagnella, P. Patruno, P. Pottier, and A. Rossa. 2000. Intercomparison and evaluation of precipitation forecasts for MAP seasons 1995 and 1996. *Meteorol. Atmos. Phys.* **72**, 111–129.

Mook, W. M. E. 2001. *Environmental Isotopes in he Hydrological Cycle. Principles and Applications.* UNESCO/IAEA Series, http://www.-naweb.iaea.org/napc/ih/volumes.asp.

Morgenstern, O., and H. C. Davies. 1999. Disruption of an upper-level PV-streamer by orographic and cloud-diabatic effects. *Contr. Atmos. Phys.* **72**, 173–186.

Mosley-Thompson, E., C. R. Readinger, P. Craigmil, L. G. Thompson, and C. A. Calder. 2005. Regional sensitivity of Greenland precipitation to NAO variability. *Geophys. Res. Lett.* **32**, L24707.

Moulin, C., C. E. Lambert, U. Dayan, V. Masson, M. Ramonet, P. Bousquet, M. Lagrand, Y. J. Balkanski, W. Guelle, B. Marticorean, G. Bergametti, and F. Dulac. 1998. Satellite climatology of African dust transport in the Mediterranean atmosphere. *J. Geophys. Res.* **103**, 13137–13144.

Newell, R. E., N. E. Newell, Y. Zhu, and C. Scott. 1992. Tropospheric rivers? - A pilot study. *Geophys. Res. Lett.* **12**, 2401–2404.

Noone, D., and I. Simmonds. 2002. Annular variations in moisture transport mechanisms and the abundance of d18O in Antarctic snow. *J. Geophys. Res.* **107**, 4742.

Numaguti, A. 1999. Origin and recycling processes of precipitating water over the Eurasian continent: Experiments using an atmospheric general circulation model. *J. Geophys.Res.* **104**, 1957–1972.

O'Dowd, C. D., M. C. Facchini, F. Cavalli, D. Ceburnis, M. Mircea, S. Decesari, S. Fuzzi, Y. J. Yoon, and J.-P. Putaud. 2004. Biogenically driven organic contribution to marine aerosol. *Nature* **431**, 676–680.

O'Reilly, J. E., S. Maritorena, B. G. Mitchell, D. A. Siegel, K. L. Carder, S. A. Garver, M. Kahru, and C. McClain. 1998. Ocean color chlorophyll algorithms for SeaWiFS. *J. Geophys. Res.* **103**, 24937–24953.

ornexl, B. E., M. Gehre, R. Hofling, and R. A. Werner. 1999. On-line delta O-18 measurement of organic and inorganic substances. *Rapid Comm. Mass Spectr.* **13**, 1685–1693.

Osada, K., H. Iida, M. Kido, K. Matsunaga, and Y. Iwasaka. 2004. Mineral dust layers in snow at Mount Tateyama, Central Japan: formation processes and characteristics. *Tellus* **56B**, 382–392.

Papayannis, A., D. Balis, V. Amiridis, G. Chourdakis, G. Tsaknakis, C. Zerefos, A. D. A. Castanho, S. Nickovic, S. Kazadzis, and J. Grabowski. 2005. Measurements of Saharan dust aerosols over the Eastern Mediterranean using elastic backscatter-Raman lidar, spectrophotometric and satellite observations in the frame of the EARLINET project. *Atmos. Chem. Phys.* **5**, 2065–2079.

Petit, J. R., J. W. C. White, N. W. Young, J. Jouzel, and Y. S. Korotkevich. 1991. Deuterium excess in recent Antarctic snow. *J. Geophys. Res.* **96**, 5113–5122.

Preunkert, S., D. Wagenbach, M. Legrand, , and C. Vincent. 2000. Col du Dôme (Mt Blanc Massif, French Alps) suitability for ice-core studies in relation with past atmospheric chemistry over Europe. *Tellus* **52B**, 993–1012.

Prodi, F., and G. Fea. 1978. Transport and deposition of Saharan dust over Alps. *Proceedings der 15. Internationalen Tagung für Alpine Meteorologie*. Vol. 40. Grindelwald. Pp. 179–182.

Prodi, F., and G. Fea. 1979. A case of transport and deposition of Saharan dust over the Italian peninsula and southern Europe. *J. Geophys. Res.* **84**, 6951–6960.

Prospero, J. M., and P. J. Lamb. 2003. African droughts and dust transport to the Caribbean: Climate change implications. *Science* **302**, 1024–1027.

Ravi, S., P. D'Odrioco, T. M. Over, and T. M. Zobeck. 2004. On the effect of air humidity on soil susceptibility to wind erosion: The case of air-dry soils. *Geophys. Res. Lett.* **31**, L09501. 10.1029/2004GL019485.

Reale, O., L. Feudale, and B. Turato. 2001. Evaporative moisture sources during a sequence of floods in the Mediterranean region. *Geophys. Res. Lett.* **28**, 2085–2088.

Robasky, F. M., and D. H. Bromwich. 1994. Greenland precipitation estimates from the atmospheric moisture budget. *Geophys. Res. Lett.* **21**, 2495–2498.

Rogers, J. C., J. F. Bolzan, and V. A. Pohjola. 1998. Atmospheric circulation variability associated with shallow-core seasonal isotopic extremes near Summit, Greenland. *J. Geophys. Res.* **103**, 11205–11219.

Rouault, M., P. Florenchie, N. Fauchereau, and C. J. C. Reason. 2003. South East tropical Atlantic warm events and southern African rainfall. *Geophys. Res. Lett.* **30**, 8009.

Rozanski, K., and C. Sonntag. 1982. Vertical distribution of deuterium in atmospheric water vapour. *Tellus* **34**, 135–141.

Rozanski, K., C. Sonntag, and K. O. Münnich. 1982. Factors controlling stable isotope composition of European precipitation. *Tellus* **34**, 142–150.

Rozanski, K., L. Arágguas-Arágguas, and R. Gonfiantini. 1993. Isotopic patterns in modern global precipitation. Pp. 1–37. In *Climate Change in Continental Isotopic Records*. S. et al. (ed.). Geophysical Monographs. American Geophysical Union.

Russell, G. L., and J. A. Lerner. 1981. A new finite-differencing scheme for the tracer transport equation. *J. App. Met.* **20**, 1483–1498.

Ryall, D. B., R. G. Derwent, A. J. Manning, A. L. Redington, J. Corden, W. Millington, P. G. Simmonds, S. O'Dotherty, N. Carslaw, and G. W. Fuller. 2002. The origin of high particulate concentrations over the United Kingdom, March 2000. *Atmos. Env.* **36**, 1363–1378.

Saurer, M., I. Robertson, R. Siegwolf, and M. Leuenberger. 1998. Oxygen isotope analysis of cellulose: An interlaboratory comparison. *Analyt. Chem.* **70**, 2074–2080.

Schär, C. 2002. *Numerische Methoden in der Umweltphysik*. IACETH, ETH Zürich.

Schär, C., D. Lüthi, U. Beyerle, and E. Heise. 1999. The soil precipitation feedback: A process study with a regional climate model. *J. Climate* **12**, 722–741.

Schär, C., P. L. Vidale, D. Lüthi, C. Frei, C. Häberli, M. A. Liniger, and C. Appenzeller. 2004. The role of increasing temperature variablility in European summer heatwaves. *Nature* **427**, 332–336.

Scherrer, S. C., M. Croci-Maspoli, C. Schwierz, and C. Appenzeller. 2006. Two-dimensional indices of atmospheric blocking and their statistical relationship with winter climate patterns in the Euro-Atlantic region. *Int. J. Climatol.* **26**, 233–249.

Schmidli, J., C. Schmutz, C. Frei, H. Wanner, and C. Schaer. 2002. Mesoscale precipitation variability in the region of the European Alps during the 20th century. *Int. J. Climatol.* **22**, 1049–1074.

Schmidt, G. A. 1999. Forward modeling of carbonate proxy data from planktonic foraminifera using oxygen isotope tracers in a global ocean model. *Palaeooceanography* **14**, 482–497.

Schoenfeldt, H.-J., and S. von Loewis. 2003. Turbulence-driven saltation in the atmospheric surface layer. *Meteorol. Z.* **12**, 257–268.

Schütz, L., and M. Sebert. 1987. Mineral aerosols and source identification. *J. Aerosol Sci.* **18**, 1–10.

Schwierz, C. 2001. *Interactions of Greenland-scale orography and extra-tropical synoptic-scale flow*. Diss. eth no. 14356. IACETH, ETH Zürich.

Schwikowski, M., P. Seibert, U. Baltensperger, and H. W. Gäggeler. 1995. A study of an outstanding Saharan dust event at the high-alpine site Jungfraujoch, Switzerland. *Atmos. Env.* **15**, 1829–1842.

Seimon, A. 2003. Improving climatic signal representation in tropical ice cores: A case study from the Quelccaya Ice Cap, Peru. *Geophys. Res. Lett.* **30**, 1772.

Seneviratne, S. I., J. S. Pal, E. A. B. Eltahir, and C. Schaer. 2002. Summer dryness in a warmer climate: a process study with a regional climate model. *Climate Dynamics* **20**, 69–85.

Shao, Y., and L. M. Leslie. 1997. Wind erosion prediction over the Australian continent. *J. Geophys. Res.* **102**, 30091–30105.

Simmonds, A. J., and J. K. Gibson. 2000. *The ERA-40 Project Plan*. ECMWF. Shinfield Park, Reading, UK.

Simmons, A. J., and D. M. Burridge. 1981. An energy and angular-momentum conserving vertical finite-difference scheme and hybrid vertical coordinates. *Mon. Wea. Rev.* **109**, 758–766.

Smolarkiewicz, P. K. 2005. Multidimensional positive defnite advection transport algorithm: An overview. *Int. J. Numer. Meth. Fluids* Pp. 1–22.

Smolarkiewicz, P. K., and L. G. Margolin. 1998. MPDATA: A finite-difference solver for geophysical flows. *J. Comput. Phys.* **140**, 459–480.

Smolarkiewicz, P. K., and W. W. Grabowski. 1990. The multidimensional positive definite advection transport algorithm - nonoscillatory option. *J. Comput. Phys.* **86**, 355–375.

Sodemann, H. 2000. *Relationships between the Origin of Air Masses and Carbon Monoxide Measurements at the Cape Point Trace Gas Monitoring Station*. B.sc. (hons) thesis. University of Cape Town, SA. Department of Environmental and Geographical Science.

Sodemann, H., A. Palmer, C. Schwierz, M. Schwikowski, and H. Wernli. 2006. The transport history of two Saharan dust events archived in an Alpine ice core. *Atmos. Chem. Phys.* **6**, 667–688.

Sodemann, H., and T. Foken. 2005. Special characteristics of the temperature structure near the surface. *Theor. Appl. Climatol.* **80**, 81–89.

Steig, E. J., P. M. Grootes, and M. Stuiver. 1994. Seasonal precipitation timing and ice core records. *Science* **266**, 1885–1886.

Steinacker, R. 1984. The isentropic vorticity and the flow oer and around the Alps. *Riv. Met. Aer.* **43**, 79–83.

Stohl, A. 1998. Computation, accuracy, and applications of trajectories - a review and bibliography. *Atmos. Env.* **32**, 947–966.

Stohl, A., and P. James. 2004. A Lagrangian analysis of the atmospheric branch of the global water cycle. Part I: Method description, validation, and demonstration for the August 2002 flooding in Central Europe. *J. Hydrometeorol.* **5**, 656– 678.

Stohl, A., and P. James. 2005. A Lagrangian analysis of the atmospheric branch of the global water cycle. Part II: Moisture transports between Earth's ocean basins and river catchments. *J. Hydrometeorol.* **6**, 961–984.

Stuiver, M., and M. Grootes. 2000. GISP-2 oxygen-isotope ratios.. *Quatern. Res.* **53**, 277–284.

Sturm, K., G. Hoffmann, B. Langmann, and W. Stichler. 2005. Simulation of d18O in precipitation by the regional. *Glob. Planet. Change* **19**, 3425–3444.

Tafferner, A. 1990. Lee cyclogenesis resulting from the combined outbreak of cold air and potential vorticity against the Alps. *Meteorol. Atmos. Phys.* **43**, 31–47.

Takacs, L. L. 1985. A two-step scheme for the advection equation with minimized dissipation and dispersion errors. *Mon. Wea. Rev.* **113**, 1050–1065.

Tegen, I., and I. Fung. 1994. Modeling of mineral dust in the atmosphere: Sources, transport, and optical thickness. *J. Geophys. Res.* **99**, 22897–22914.

Temperton, C. 1988. Implicit normal mode initialization. *Mon. Wea. Rev.* **116**, 1013–1033.

Tiedtke, M. 1989. A comprehensive mass flux scheme for cumulus parameterization in large-scale models. *Mon. Wea. Rev.* **117**, 1779–1800.

Trenberth, K. E. 1997. Using atmospheric budgets as a constraint on surface fluxes. *J. Climate* **10**, 2796–2809.

Troen, I., and L. Mahrt. 1986. A simple model of the atmospheric boundary-layer - Sensitivity to surface evaporation. *Bound.-Layer Meteorol.* **37**, 129–148.

Ulbrich, U., T. Bruecher, A. H. Fink, G. C. Leckebusch, A. Krueger, and J. G. Pinto. 2003a. The central European floods of August 2002: Part 1 - Rainfall periods and flood development. *Weather* **58**, 371–377.

Ulbrich, U., T. Bruecher, A. H. Fink, G. C. Leckebusch, A. Krueger, and J. G. Pinto. 2003b. The central European floods of August 2002: Part 2 - Synoptic causes and considerations with respect to climatic change. *Weather* **58**, 434–442.

van Bebber, W. J. 1891. Die Zugstrassen der barometrischen Minima nach den Bahnenkarten der Deutschen Seewarte von 1987–1890. *Meteorol. Z.* **8**, 361–366.

van Loon, H., and J. C. Rogers. 1978. Seesaw in winter temperatures between Greenland and Northern Europe. 1. General description. *Mon. Wea. Rev.* **106**, 296–310.

Vidale, P. L., D. Luethi, C. Frei, S. I. Seneviratne, and C. Schär. 2003. Predictability and uncertainty in a regional climate model. *J. Geophys. Res.* **108**, 4586.

Vidale, P. L., D. Lüthi, P. Heck, C. Schär, and C. Frei. 2002. *The Development of Climate HRM (ex-DWD EM)*. IACETH, ETH Zürich.

Vinther, B. M., S. J. Johnsen, K. K. Andersen, H. B. Clausen, and A. W. Hansen. 2003. NAO signal recorded in the stable isotopes of Greenland ice cores. *Geophys. Res. Lett.* **30**, 1387. 10.1029/2002GL016193.

Wagenbach, D., and K. Geis. 1989. The mineral dust record in a high altitude Alpine glacier (Colle Gnifetti, Swiss Alps). Pp. 543–564. In *Paleoclimatology and Paleometeorology: Modern and Past Patterns of Global Atmospheric Transport*. M. Leinen, and R. Sarnthein. (Eds.). Kluwer Academic Publishers.

Walker, G. T., and E. W. Bliss. 1928. World weather IV: Some applications to seasonal foreshadowing. *Mem. Roy. Meteorol. Soc.* **3**, 81–95.

Werner, M., M. Heimann, and G. Hoffman. 2001. Isotopic composition and origin of polar precipitation in present and glacial climate simulations. *Tellus* **53B**, 53–71.

Wernli, H. 1995. *Lagrangian perspectives of extratropical cyclogenesis*. Diss. eth no. 11016. IACETH, ETH Zürich.

Wernli, H. 1997. A Lagrangian-based analysis of extratropical cyclones. II: A detailed case-study. *Q. J. R. Meteorol. Soc.* **123**, 1677–1706.

Wernli, H., and H. C. Davies. 1997. A Lagrangian-based analysis of extratropical cyclones. I: The method and some applications. *Q. J. R. Meteorol. Soc.* **123**, 467–489.

Wernli, H., S. Dirren, M. Liniger, and M. Zillig. 2002. Dynamical aspects of the life cycle of the winter storm 'Lothar'. *Q. J. R. Meteorol. Soc.* **128**, 405–429.

White, J. W. C., L. K. Barlow, D. Fisher, P. Grootes, J. Jouzel, S. J. Johnsen, M. Stuiver, and H. Clausen. 1997. The climate signal in the stable isotopes of snow from Summit, Greenland: Results of comparisons with modern climate observations. *J. Geophys. Res.* **102**, 26425–26439.

Yoshimura, K., T. Oki, and K. Ichiyanagi. 2004. Evaluation of two-dimensional atmospheric water circulation fields in reanalyses by using precipitation isotopes databases. *J. Geophys. Res.* **109**, D20109.

Yoshimura, K., T. Oki, N. Ohte, and S. Kanae. 2003. A quantitative analysis of short-term 18O variability with a Rayleigh-type isotope circulation model. *J. Geophys. Res.* **108**, 4647.

Zängl, G. 2004. Numerical simulations of the 12-13 August 2002 flooding event in eastern Germany. *Q. J. R. Meteorol. Soc.* **130**, 1921 – 1940.

Acknowledgements

I would like to thank

Heini Wernli, who gave me the opportunity to dive into the complexities
of the atmospheric water cycle,

Conny Schwierz, who gave this work momentum with her flow of ideas,

Valérie Masson-Delmotte, who guided me around the crevasses of
stable isotope modelling,

Huw C. Davies, who kept things running smoothly.

Gratefully acknowledged are MeteoSwiss for access to the ECMWF ERA40 reanalysis data, Bo Vinther for seasonal isotope data from the Alfabet cores, and Martin Werner and the SWING project for their isotope GCM data. Piotr Smolarkiewicz kindly provided MPDATA advection code. Dani Lüthi provided the CHRM code and invaluable technical support.

Furthermore, many people contributed during interesting discussions to this work. In particular, I would like to thank Piotr Smolarkiewicz, Bo Vinther, Max Kelley, Michiel Helsen, Michael Bosilovich, Peter Bechtold, Adrian Tompkins, and Timothy Hall. Nele Meckler helped improving the readability of several parts of this work.

The atmosphere at the IACETH, in the dynamics group, and in our office(s) has always been very positive and friendly: thanks to all my colleagues and friends for making that arduous time a good one as well.

Finally, many thanks go to my parents, my brother, and Nele for their interest, support, and patience.

Curriculum vitae

Harald Sodemann
Born on the 16th December 1975 in Esslingen am Neckar
German citizenship

Education

2003 - 2006	PhD student at the Institute for Atmospheric and Climate Science (IAC) at the Swiss Federal Institute of Technology (ETH) Zurich under supervision of Prof. Huw C. Davies
2002	Internship at the IOC/UNESCO, Paris, France.
1996 - 2002	Studies of Geoecology at the University of Bayreuth, Germany. Specialisation in the field of Micrometeorology. Diploma thesis: "Evaluation of a parameterisation for turbulent fluxes of momentum and heat in stably stratified surface layers"
2001	Internship at the Federal Environmental Institute, Berlin, Germany.
2000	Studies in Atmospheric Sciences at the University of Cape Town, South Africa. Honours thesis: "Relationships between the origin of air masses and carbon monoxide trace gas measurements at the Cape Point trace gas monitoring station"
1995 - 1996	Work with mentally handicapped people in Dettingen/Erms, Germany
Juni 1995	Allgemeine Hochschulreife (Abitur)
1986 - 1995	Georgii-Gymnasium, Esslingen am Neckar

Conferences

Workshop on Isotope Effects in Evaporation, Pisa, Italy, May 2006
3rd General Assembly EGU, Vienna, Austria, April 2006
4th NCCR Summerschool, Grindelwald, Switzerland, September 2005
IAMAS 2005, Beijing, China, August 2005
4th Young Researchers Meeting, Gwatt, Switzerland, June 2005
2nd General Assembly EGU, Vienna, Austria, April 2005
DACH Conference, Karlsruhe, Germany, September 2004
3th NCCR Summerschool, Ascona, Switzerland, September 2004
Grand Combin Summer School, Valle d'Aosta, Italy, June 2004
3rd Young Researchers Meeting, Gwatt, Switzerland, June 2004
AGU Spring Meeting, Montreal, Canada, May 2004
1st General Assembly EGU, Nice, France, April 2004
2nd NCCR Summerschool, Grindelwald, Switzerland, September 2003
2nd Young Researchers Meeting, Münchenwiler, Switzerland, June 2003
ECMWF Training Course, Reading, April 2003
EUROTRAC/TOR-2 Workshop, Moscow, Russia, September 2002
15th AMS Boundary Layer Symposium, Wageningen, The Netherlands, July 2002
XXVII General Assembly EGS, Nice, France, April 2002
EUROTRAC-2 Conference, Garmisch-Partenkirchen, Germany, March 2002
SASAS Conference, Pretoria, South Africa, October 2000
EU Advanced Study Course, Alghero, Italy, October 1999

Publications

Sodemann, H., Palmer, A. S., Schwierz, C., Schwikowski, M. and Wernli, H., The transport history of two Saharan dust events archived in an Alpine ice core, *Atmos. Chem. Phys*, 6:667–688, 2006.

Sodemann, H. and Foken, T., Special characteristics of the temperature structure near the surface, *Theor. Appl. Climatol.*, 80:81–89, 2005.

Sodemann, H. and Foken, T., Empirical evaluation of an extended similarity theory for the stably stratified atmospheric surface layer, *Q. J. R. Meteorol. Soc.*, 130:2665–2672, 2004.

Sodemann, H., Schwierz, C., and Wernli, H.: Inter-annual variability of Greenland winter precipitation sources. Part I: Lagrangian moisture source diagnostic and North Atlantic Oscillation influence, *in prep.*

Sodemann, H., Masson-Delmote, V., Schwierz, C., Vinther, B. and Wernli, H.: Interannual variability of Greenland winter precipitation sources. Part II: North Atlantic Oscillation variability of stable isotopes in precipitation, *in prep.*

Cover photograph:
Solidified mud flow in the Swakop river, Namibia, after heavy rainfalls in early 2000.